LATITUDE
&THE MAGNETIC EARTH

STEPHEN PUMFREY

Series Editor: Jon Turney

REVOLUTIONS IN SCIENCE
Published by Icon Books UK

Originally published in 2002 by Icon Books Ltd.

This edition published in the UK in 2003
by Icon Books Ltd., Grange Road,
Duxford, Cambridge CB2 4QF
e-mail: info@iconbooks.co.uk
www.iconbooks.co.uk

Published in the USA in 2003
by Totem Books
Inquiries to: Icon Books Ltd.,
Grange Road, Duxford,
Cambridge CB2 4QF, UK

Sold in the UK, Europe, South Africa
and Asia by Faber and Faber Ltd.,
3 Queen Square, London WC1N 3AU
or their agents

Distributed to the trade in the USA by
National Book Network Inc.,
4720 Boston Way, Lanham,
Maryland 20706

Distributed in the UK, Europe, South Africa
and Asia by TBS Ltd., Frating Distribution
Centre, Colchester Road, Frating Green,
Colchester CO7 7DW

Distributed in Canada by
Penguin Books Canada,
10 Alcorn Avenue, Suite 300,
Toronto, Ontario M4V 3B2

Published in Australia in 2003
by Allen & Unwin Pty. Ltd.,
PO Box 8500, 83 Alexander Street,
Crows Nest, NSW 2065

ISBN 1 84046 486 0

Text copyright © 2002 Stephen Pumfrey

The author has asserted his moral rights.

Series editor: Jon Turney

Originating editor: Simon Flynn

Typesetting by Hands Fotoset

Printed and bound in the UK by Cox & Wyman Ltd., Reading

Cover image: Detail from 'Dr William Gilbert Showing His Experiment on
Electricity to Queen Elizabeth I and Her Court' by Arthur Ackland Hunt
(fl. 1863–1913), Bridgeman Art Library

CONTENTS

LIST OF ILLUSTRATIONS

ACKNOWLEDGEMENTS

I want to acknowledge John Schuster, whose inspirational lectures nurtured my interest in Gilbert, the late Charles Schmitt, for showing me the virtues of Renaissance Aristotelianism(s), Duane H. D. Roller, the creator of modern Gilbert studies, and Art Jonkers, from whose thesis I learned much more about magnetic navigation. Thanks also to my colleagues in the Lancaster University Department of History, especially John Brooke, who have made my academic career more enjoyable than William Gilbert's.

I owe deep gratitude to the Icon Books team, especially the patient Simon Flynn, the incisive Jon Turney, and the efficient Alison Foskett, who have made this a better, shorter book. Thanks also to Art Jonkers, Ian Stewart and Richard Cunningham for making time to read a draft. I cannot put any remaining errors down to my instruments. Finally, I want to thank my parents for their support, my wife Dee for her example, and my two little spheres of activity for giving me something other than magnetism to think about. *Latitude & the Magnetic Earth* is therefore dedicated to Niall and Caitríona.

ABOUT THE AUTHOR

Stephen Pumfrey is a senior lecturer in the history of science at Lancaster University. His early life shadowed Gilbert's, from Gilbert Road and St John's College in Cambridge, to London. He now lives in Manchester with his wife, also an academic, and two small children. He enjoys walking, not sailing, and his only serious attempt at compass navigation ended in failure.

INTRODUCTION:
THE MAGNETIC REVOLUTION

We live in a magnetic world. My refrigerator has magnets in its motors, thermostat and door lock, and decorative magnets on the door. The monitor, disks and drives of the computer on which I'm typing depend on magnets. So do electric switches and valves, every Walkman and telephone, every electric motor and loudspeaker, not to mention CAT scanners and particle accelerators. Over 30 grams of magnets are manufactured every year for every person on Earth, and a rich Westerner will own hundreds. The Earth itself is a giant natural magnet. I can navigate its surface using a magnetic compass. Astrophysicists have measured the field strength of other planets and even interstellar space. As I write, the Sun is reaching the height of the eleven-year cycle of its magnetic sunspots. A solar flare might distort the Earth's magnetic field, rendering my compass unreliable and knocking out communications systems. As the twenty-first century develops, mag-lev trains travelling at 800 kph could replace polluting aeroplanes. By then, however, radiation from mobile phones may have fried our brains, for every atom in the universe, including those in human tissue, is basically magnetic. As Dick Tracy's sidekick, Diet Smith, said in the 1960s comics, 'He who controls magnetism controls the world.' The name of the scientist who created the magnetic world, whose name once ranked alongside those of Copernicus or Galileo, was the Elizabethan doctor, William Gilbert.

We don't have to go very far back in time to leave our magnetic universe behind. Not much more than 150 years ago, Michael Faraday discovered electromagnetic induction and, with it, the principles of the electric motors and dynamos that accelerated the Industrial Revolution. He also developed the concepts of a magnetic field and lines of force. It was only in 1820 that Hans-Christian Oersted announced that the forces of magnetism and electricity were connected. Before that, you could only find magnets in nautical compasses, gentlemen's science kits and occasionally in the naturally occurring iron ore called magnetite or lodestone.

Going back to the eighteenth century, magnetic and electrical apparatus were prized components of the science kits that Europe's gentry bought as fashion items. They were fascinated by the strange, even magical, forces of attraction and repulsion that we experience when we bring two magnets together, or that make a balloon stick on the wall using static electricity. The eighteenth century's scientific hero, Sir Isaac Newton, had declared magnetism to be, like gravity and electricity, one of the fundamental forces of God's creation. What could be a more edifying leisure activity than to conjure up the divine spark? People were convinced of the similarities between electrical and magnetic phenomena. Luigi Galvani famously 'galvanised' a frog's leg into motion by running an electric current through the muscle's nerve. Not long after Galvani's announcement of 'animal electricity', Franz Mesmer claimed to have discovered and bottled 'animal magnetism'. He briefly made a good living in Paris magnetically curing gentlemen and gentlewomen sufferers (he preferred gentlewomen) of their pain.

Newton published his masterwork, *Philosophiae Naturalis Principia Mathematica* [*The Mathematical Principles of Natural Philosophy*], in London a century earlier in 1687. If we go back in London's history another 100 years, the idea that the Earth itself was a magnet had yet to be heard. Yet London in 1587 was in the middle of its first boom in magnet making – for nautical compasses. The port of London was home to England's expanding mercantile and naval fleets. Elizabethan England was a fast developing nation. It was transforming from Europe's economic and cultural backwater into a minor imperial power that could compete with Spain and France. The Spanish Armada of 1588 was expected, as was its defeat. England's new strength and self-confidence owed much to its sailors and their expertise. In particular, their compasses had recently become as good as any in the world. But if the new heroic English navigators like Francis Drake could tell magnetic north from south, they had no better idea than did Spanish experts how the compass worked. In 1587 no one had even a suspicion that the magnetic compass worked by aligning itself with the magnetic force of planet Earth. That deceptively simple theory, so familiar to us, was almost unthinkable. But, some time in the 1580s, the unknown Dr Gilbert began to think the unthinkable, and to invent the magnetic universe.

Gilbert blazed into the scientific world like a comet. Before 1600, the year he published his only book, *De Magnete*, or *On the Lodestone*, few people outside his

London medical circle had heard of him. But *De Magnete* was a sensational work that turned him into the first modern English scientist of international repute. Its impact was not just due to Gilbert's extraordinary new theory and its applications. There was also the novel way that he used and described 'experiments that appeal plainly to the senses' to prove it. Then, as now, scientists were expected to show how their work built on respected, published conclusions. Gilbert, with a revolutionary's lack of respect and caution, declared in his preface to *De Magnete* that the world was already:

> [full] *of books of the more stupid sort whereby the common herd . . . are led to profess themselves philosophers, physicians, mathematicians, and astrologers, the while ignoring and contemning men of learning . . . why should I submit this noble and (as comprising many things before unheard of) this new and inadmissible philosophy to the judgement of men who have taken oath to follow the opinions of others, to the most senseless corrupters of the arts, to lettered clowns, grammatists, sophists, spouters, and the wrong-headed rabble, to be denounced, torn to tatters and heaped with contumely. To you alone, true philosophers, ingenuous minds, who not only in books but in things themselves look for knowledge, have I dedicated these foundations of magnetic science – a new style of philosophizing.*

This new style meant that, for an unknown physician from Colchester whose debut discussed an obscure, puzzling mineral in a print run of a few hundred copies, Gilbert's star began to burn brightly. In Venice in 1602,

Galileo was excited, and his friend Giovanni Sagredo wrote to Gilbert of their enthusiasm. By 1603, the German astronomer Johann Kepler wished that he 'had wings, that I might fly to England to talk with Gilbert'. Kepler's mission to Europe's intellectual backwater would have been fruitless, because Gilbert's light was extinguished by the plague in 1603. Sagredo's letter remained his only tangible evidence that *De Magnete* would have the influence he hoped for.

By the 1630s, though, the revolutionary impact of Gilbert's ideas was clear. Kepler cited them as a foundation of his elliptical astronomy. Galileo's use of them was presented as damning evidence in his trial for heresy. Catholic scientists of the Society of Jesus had begun a concerted effort to discredit them. Ocean navigators, including Jesuit missionaries, now carried novel instruments described in *De Magnete*, believing that Gilbert's ideas offered a solution to the problem of determining both latitude and longitude at sea. Magnetic scientists scrambled to win vast state prizes. From then until now, Gilbert has been remembered as the founder of a science of magnetism, although the enthusiasm of astronomers and navigators that first brought him international fame has long dissipated.

Today we can see that Gilbert revolutionised Earth science, and the Earth itself, in three ways. The first, most basic way was his brilliant conceptualisation of our Earth as a giant spherical magnet. His geomagnetic theory and his experimental proofs of it form his enduring legacy to modern science. From this one simple and completely novel insight, Gilbert not only provided a rational explanation of the previously baffling behaviour of magnetic compass needles, he also discovered a new way for sailors to find their latitude, and inspired numerous attempts to solve the problem of longitude magnetically.

Second, Gilbert literally revolutionised the Earth, because he was one of the handful of scientists who, by 1600, had committed themselves to the cosmological principles of Copernicus. In that small group of believers in the Earth's daily and annual rotation, which included the young Kepler and Galileo, Gilbert was unique, because he alone offered a physical cause for the Earth's motion. That cause was the magnetic force that planet Earth possessed. It was Gilbert's revolutionary Copernican physics that got him a bad name with Jesuits and Inquisitors. Of course, Isaac Newton's laws of motion and gravitational attraction successfully replaced Gilbert's magnetic Copernicanism. For some historians this adds piquancy to Gilbert's story – how could such a modern mind get things so wrong?

The piquancy is sharpened by the third strand of Gilbert's revolution. *De Magnete* has, with some justification, been called the first thoroughly experimental treatise. Gilbert himself was bold enough to think so. Throughout the work he contrasted his method, of taking nothing on authority, of starting from tried and tested empirical data, of devising careful laboratory experiments and of describing them in sufficient detail to allow others to replicate them, with that of his predecessors. He criticised them mercilessly for their uncritical acceptance of established authorities such as the ancient Greek Aristotle, and their credulous repetition of myths or hearsay as empirical facts, such as the

6

power of garlic oil to weaken a magnet, or even the power of a lodestone to detect adultery. Yet he also suggested that the Earth was animated by, or alive with, magnetic force.

Gilbert's revolution, then, is much more complicated, and more interesting, than the triumph of modern empirical science. Even if Gilbert had simply 'discovered' geomagnetism as an 'empirical fact', which he obviously did not, the scientific world of his time could not have accepted geomagnetism as an interesting addition to human knowledge of the world. The long-established and dominant natural philosophy of Gilbert's era derived from Aristotle. Aristotelian matter theorists had a full explanation of what constituted both the Earth as a body and 'earth' as an element. The fundamental principles of their theory precluded any ideas of the Earth having an attractive power to move other matter magnetically, or to move itself. Gilbert was perfectly aware of that. He thought that if he could prove the existence of the Earth's magnetism, then he would have refuted the very core of Aristotelian science. A new natural philosophy would be needed, and Gilbert was ready with the first new science of the modern age – 'magnetic philosophy'. The Essex doctor did not precipitate a scientific revolution with the weight of his magnetic facts; he plotted the revolution before he had them!

PART I

BEFORE GILBERT'S REVOLUTION

· CHAPTER 1 ·

WHO WAS WILLIAM GILBERT?

Strictly speaking, William Gilbert was a natural philosopher, not a scientist. Our modern sciences evolved out of natural philosophy. They are much more integrated with mathematics and technology, and much more separated from general philosophy and theology. Gilbert's work, especially his respectful use of navigational expertise, played a part in that evolution, but he was still basically a Renaissance natural philosopher – interested in the whole range of causes that God had used to make our orderly universe. With due historical respect to his own identity, we will call him a natural philosopher, and the subject that he created by his own phrase of 'magnetic philosophy'.

Few natural philosophers, let alone scientists, have left so few traces of their lives and of the development of their ideas. Ever since 1600, when *De Magnete* began to stun the likes of Johann Kepler and Galileo, people have tried to find out who was this 'William Gilbert of Colchester, Physician of London', as the title page described him. Who could create a new science of the Earth, demonstrate the power of experimental investigation and dare to prove that the Earth moved? No natural philosophy of international repute had ever come out of London, and Colchester was on few of the maps owned by Europe's savants.

Illustration 1: Portrait of William Gilbert (1544–1603).
We know Gilbert's face as uncertainly as other facets of
his life. This late eighteenth-century engraving is the only
reliable portrait, showing him in his late forties or fifties.
It was made from an original that is now lost, which may

Some things have changed. Colchester now trades off Gilbert's fame. His family home, Tymperley's, is a museum, the Town Hall is adorned with pictures (including the original of the cover of this book) and a statue, and a school is named after him. But Gilbert remains an enigma. If only he had published a series of books from his youth to his dotage, as Kepler did. But there is only *De Magnete*. If only he had left sheaves of manuscripts, letters and apparatus, like Galileo. But most of his effects were lost in the Great Fire of London. If only he had been a member of a scientific society or correspondence network, like Robert Boyle or Rene Descartes, but he was born fifty years too early. If only an explicit, gossipy diary had surfaced, like Robert Hooke's. If only he had survived the outbreak of plague in 1603 to debate and explain his ideas.

The enigmatic background of Elizabeth I's scientific genius has added to his mystique. Earlier historians left him as just that – a genius who mysteriously transcended his Elizabethan age and anticipated modern, experimental science. *De Magnete* has been used as an example of every passing orthodoxy of what makes good science. Gilbert has appeared as an inductivist who stuck, as it were, to 'the facts, just the facts' of experimental magnetism; or a materialist who turned the practical methods of craftsmen into science; even a hypothetico-deductivist

have been the same one seen in Oxford University in the mid-seventeenth century. Gilbert's hand rests on a globe, symbolising his interest in the Earth and navigation. Compare it with the portrait of Francis Bacon (Illustration 14).

who devised experiments to test bold, unproven hypotheses. It is a sign of the rich complexity of his work that he has been a scientist for all seasons.

Biographically, there have been plenty of attempts to turn the bare facts into likely stories. Among the myths that inflate Gilbert's greatness are that he came from a long line of gentlemen, left England in search of enlightenment, took his MD in Italy, met Galileo, stayed single so as to devote himself to science, and that Queen Elizabeth personally financed his research. Fortunately, even when we strip all these myths away, enough remains to get an insight into the creation of a natural philosophical revolutionary.

William Gilbert of Colchester and Cambridge

William Gilbert came from a solid family of East Anglian gentry. Although Gilbert relied on, and praised, the work of navigators, instrument makers and metal-workers, only well-to-do authors like him got a proper hearing. Elizabethan society was a hierarchical one. It would be another 100 years before someone of genuine artisan stock was acceptable as a natural philosopher, and almost as long before an aristocrat would think it dignified to get his hands dirty with experimentation. Gilbert's insistence that artisans had a better understanding of nature than university philosophers has just a whiff of social radicalism about it.

The cloth trade was an important part of the East Anglian economy, and Gilbert's paternal grandfather was a Suffolk weaver, prominent enough to be made 'Sewer to the Chamber' under Henry VIII. The Gilberts benefited from the considerable expansion of the gentry in late Tudor England. There was more wealth, more places at Oxbridge, more professional positions, and more routes to the centres of power in the royal court and London generally. Gilbert and his fellow courtier and scientific reformer, the lawyer Francis Bacon, both seized the new opportunities. Gilbert's father, Jerome Gilberd (as his family most commonly wrote the name) cashed in on his father's wealth by getting a legal education, and moving the family to Colchester, where he became the town's Recorder, or minor judge.

Jerome's major contribution was to populate the Essex–Suffolk border regions with healthy Gilberds. William was the first of four children from his first marriage, and he had nine more with his second wife, the seemingly well-to-do Jane Wingfield. Defying the infant mortality rates of the time, eight of the nine children lived long enough to marry. Gilbert remained single, and was close to his family, especially to his half-brothers Ambrose and William junior.

Ambrose and William erected an inscribed monument to their famous brother in Holy Trinity Church, Colchester. It says that he 'died in the year of Human Redemption 1603, on the last day of November, in the 63rd year of his age', which explains why many authorities have assumed that Gilbert was born in 1540. But even this is almost certainly wrong – the more likely date is 1544. Someone cast Gilbert's horoscope in later life, and used the time and date of 2.20 pm on 24 May 1544. The year 1544 fits better with other events in his life but, with no surviving baptismal record, even the first biographical fact is missing.

As the first-born son of a rising professional, Gilbert got the education he needed to consolidate the family's position. He arrived into the care of the schoolmen of St John's College, Cambridge in 1558 which, if he was born in 1544, was at the customary age of fourteen. The Cambridge of Gilbert's era was a small, very traditional, frankly backward place compared with vibrant university towns and cities like Paris, Padua or Protestant Basle. Despite reforming forces like John Caius, medical education was especially ossified. Intellectually ambitious physicians left for foreign faculties, especially Padua, where the graduate of Caius' College, William Harvey, enrolled in 1600. After his fame, it was widely assumed that Gilbert had done something similar, but college records show that, in fact, he stayed at home. He got his BA in 1561, his MA in 1564 and a Cambridge MD in 1569. He succeeded well enough to be elected a fellow of St John's in 1561. He held the College's junior position of mathematical examiner in 1565 and 1566, and was its bursar in 1570. A career in academia, lecturing in medicine perhaps, lay open, but Gilbert turned his back on Cambridge. Everything he wrote subsequently suggests that he came to detest the kind of scholarship and natural philosophy that Cambridge epitomised.

University education in Cambridge was extremely conservative. The aim was to drum into BA students the principles of the great classical authorities – Aristotle in philosophy, Ptolemy in astronomy and Galen in medicine – and their more recent commentators. Although inexpensive printed texts now existed, the medieval practice of 'lecturers' repeating aloud works for students to copy and learn persisted. Statutes outlined fines for scholars who criticised the approved authorities.

Nevertheless, by Gilbert's time Cambridge had adopted some of the reforms pioneered by Italian universities in the Renaissance. There was more emphasis upon original Greek texts, which Gilbert probably read. Increased availability of modern and recovered ancient texts broadened the range of sources. Aristotle was still the authority in the discipline of natural philosophy. Teaching also became more topic-centred, with less value placed on dissecting Aristotle's books line by line. But magnetism was barely mentioned, let alone a topic of enquiry.

In his foundation year in the junior arts faculty, Gilbert would have brushed up on the 'trivial' subjects of grammar, rhetoric and dialectic, and begun the mathematical quadrivium of arithmetic, geometry, astronomy and music. Natural philosophy came in his second year. It was still 150 years away from becoming a central university discipline, but it was more important than mathematics, because it dealt with the facts and causes of the natural world. Mathematics was presumed able to describe but not to explain phenomena such as the motions of the planets. Knowledge of natural causes was more useful to the aspiring cleric, doctor or educated bureaucrat. The final BA year was more serious, and covered ethics, metaphysics and theology. Gilbert would then have been given an expanded reading list, which he would have been tested on when he applied for his MA.

Now he needed a profession. The higher faculties of theology, law and medicine offered three. Judging by Gilbert's later criticism of theologians, he was not cut out for a career in the Church, and medicine clearly held more appeal than his father's profession of law. Medicine, or physic, was closely linked to natural philosophy. Indeed, the name physic derives from *physica*, a Latin synonym for natural philosophy, and physicians were supposed to be distinguished by their knowledge of the 'physics' or science of the human body. Perhaps Gilbert's training in medical theory inaugurated his dissatisfaction with the conventional doctrines of matter, especially those of the four elements and their origin in the four qualities of heat, cold, moistness and dryness.

When Gilbert was elected a fellow, his teaching duties probably included 'lecturing' Aristotle's natural philosophy texts, such as *On the Heavens* and *Meteorology* – the only two books that we know Gilbert owned. The demands of mathematical examining were not onerous, given that mathematics was the first year, Cinderella subject of the curriculum. Whatever competence he acquired, Gilbert's colleagues did not rate his grasp of astronomy, the highest branch of mathematics. This is not surprising, for his conservative views on the relative status of mathematics and natural philosophy came straight out of the traditional academy. What is surprising is that a Copernican should have been so dismissive.

Despite his conservative view that mathematics did not explain anything, at some point Gilbert began to question the orthodox natural philosophy of his day. There is evidence that Gilbert began seriously to doubt some of Aristotle's theories in the late 1560s and undertook some independent research. The focus of his doubt was Aristotle's *Meteorology*. Aristotelian meteorology was the science of the 'sphere' of air. Like modern meteorology, it dealt with atmospheric phenomena arising from air, its movements and substances in it, such as water vapour or fiery matter. It used Aristotle's matter theory to explain winds, clouds, rainbows and other appearances in the sky, such as comets and meteors.

Gilbert became interested in connections between the weather and the position and influence of the planets. This was part of a Galenic doctor's training. Doctors were concerned to treat a patient when the qualities ascribed to each planet best matched the balance of qualities in the patient's temperament. Gilbert's earliest work was a notebook of dated records of the weather, correlated with astrological data. He kept his own from around 1569. He concluded that Aristotle's explanations were wrong, and a new meteorology was needed. His revolt had begun. By the 1580s, Gilbert had drafted 'A New Meteorology in opposition to Aristotle', which criticised the doctrines of elements and qualities.

Dr Gilbert, London Physician

Ten years earlier, Gilbert had made his major professional move. In the 1570s, Cambridge and Oxford colleges were full of bright young men on the lookout for more exciting prospects in London. One reason for this was the intellectual current of civic humanism. Some progressive Elizabethans wanted to catch up with the European Renaissance. They advocated the *vita activa*, the active life of putting one's talents at the disposal of powerful people in government and commerce, for the greater benefit of the state, and for the greater status and rewards. They came to despise the *vita contemplativa*, epitomised by conservative academics, who claimed to have knowledge of everything but who could apply it to nothing useful. Oxbridge values, the reverence for classical authorities, the minutiae of scholastic debate, the pettiness and detachment of college life, were of little value in the new centre of action – London and the court.

So Gilbert joined the brain drain to London, certainly to set up as a doctor, and, maybe, to find the freedom, leisure and resources to continue his intellectual revolt against Aristotle. In mainland Europe many famous works of natural philosophy were published by learned court physicians, but no English doctor before Gilbert had made an impact.

His life in the 1570s is a complete blank. His brothers' inscription is the only evidence that he was practising in London. However, by 1577 he was well-to-do enough to get a grant of arms from the College of Heralds, proof (often spurious, but always paid for) of established gentility. And by 1580 he re-appears as the complete and successful doctor.

London held the richest pickings for a physician, but they were carved up by the august College (now Royal College) of Physicians. It had been set up under Henry VIII to regulate the crowd of trained physicians, self-styled doctors, apothecaries, barber-surgeons, wise-women, peddlers of patent remedies and outright charlatans who competed to cure the city's hundreds of thousands of souls and their bodies. The top men were the mere forty physicians who were not just licensed by the College but were elected to its Fellowship. Even the College's superb records cannot tell us when Gilbert joined, but by 1581 he had begun nine years' service as a College censor. Censors were the London medical police, checking on qualifications and standards of practice, dragging wise-women before their court, fining apothecaries for trespassing into the physicians' domain of diagnosis and presenting would-be physicians to a panel of examiners.

To reach such eminence, he needed important patrons. Nobody, from the humblest supplier of nautical instruments, through doctors and clergy, to the highest nobles, attained any position, power or wealth without their support. As an aspiring physician, Gilbert needed them even more. A doctor's work, taking urine, examining syphilitic sores, advising on personal diet and hygiene, made him intimate with a statesman's body and household. Personal recommendation of a physician's ability, bedside manner and discretion were crucial. And nobles often trusted physicians with their secrets.

And so, in January 1582, we find Gilbert on a mission for the Earl of Shrewsbury to a client.

I have thought it good to send the letter [about finance] *unto you, which I would have you keep safe. This gentleman, Doctor Gilbert, was sent from her Majesty by my lord of Leicester's means. ... I have spoken to him about some biscuit bread, which is not made by common poticars, and also a serecloth, to use for my gout, which he has promised to send. See him well recompensed, for surely, for the small talk I have had with him, I have found him a sensible man; therefore seek to be acquainted with him, and be very friendly of him.*

21

Gilbert the physician had made it. The next eighteen years, leading up to his *annus mirabilis* of 1600, saw success after success. He moved smoothly up the College of Physician's hierarchy, holding all the key offices, and being rewarded with the presidency in 1600. He had still not published anything, so most college fellows, apart from his doctor friends, were probably ignorant of his dismissive attitudes to Galen's authority.

At the same time, his network of noble patrons expanded to include the very best, William Cecil, Lord Burghley, Elizabeth's most trusted politician. Gilbert ministered to the dying Cecil and his wife, and continued to work for his son Robert, whom Cecil groomed as his political successor. But if Gilbert tried to get his patrons interested in magnetic philosophy, then he failed. Like doctors, learned exponents of unconventional natural philosophy needed support from the rich and powerful. A noble patron would find them a position, guarantee their leisure time, support their research, probably pay for publication, and protect them from criticism. In return, they would get an obsequious letter of dedication praising their judgement, and the reflected glory of backing an intellectual winner.

Very unusually, *De Magnete* had no such dedication, and no mention of patrons, except in a scathing reference to:

[those authors] *who publish things not even worthy of record; who, pilfering some book, grasp for themselves from other authors, and go a-begging for some patron, or go a-fishing among the inexperienced and young for a reputation; who seem to transmit from hand to hand,*

as it were, erroneous teachings in every science and out
of their own store now and again to add somewhat
of error.

Are these the words of a bitter man who found Elizabeth's courtiers too narrow-minded to associate themselves with magnetic philosophy, or a real radical who refused to tailor his ideas to any value system, scholastic or humanist? Gilbert seemed to revel in the idea that his work was too hot to handle. In any event, it adds to the impression that Gilbert led a double life – a social conformist, but intellectual revolutionary.

Gilbert and the Navigators

Just before his attack on patron-seeking scholars, Gilbert listed people he thought truly 'learned men'. They were:

that most accomplished scholar Thomas Hariot, Robert
Hues, Edward Wright, Abraham Kendall, all English-
men; others have invented and published magnetic
instruments and ready methods of observing, necessary
for mariners and those who make long voyages: as
William Borough in his little work the Variation of
the Compass, *William Barlo[w] in his* Supplement,
Robert Norman in his New Attractive *– the same Robert*
Norman, skilled navigator and ingenious artificer, who
first discovered the dip of the magnetic needle.

These names show how Gilbert looked beyond the intellectual world of physicians and natural philosophers for inspiration and information. They are the cream of

Queen Elizabeth I's navigation experts and mathematical practitioners.

Gilbert probably had his big idea of the magnetic Earth early in the 1580s, but it was just an idea, supported by little evidence. The best people to confirm it, those with the best knowledge of magnetism, were these compass experts. In March 1588, Gilbert's medical expertise gave him the perfect introduction. England expected Philip of Spain's invading Armada to sail any day. Elizabeth's navy was ready, but its sailors were hit by disease. Her Privy Council sought help from the College of Physicians, and:

> *For remedie of the dyseased and for staie of further contagion their Lordships thought meet that some lerned and skillfull phisicions shoud presently be sent thither; and for that their Lordships heard that good reporte of the sufficiency, learninge and care of Dr Gilbert, Dr Marbeck, Dr Browne and Dr Wilkinson, as they were thought very fytt persons to be employed in the said Navye to have care of the helthe of the noblemen, gentlemen and others in that service.*

The College was to choose two, who would get whatever 'entertainment as should be to their contentment', provided that they carried 'with them a convenyent quantytie of all soche drogues as should be fyt for medicine and cure'.

Did Gilbert go? There's no firm evidence, but he did develop remarkably good contacts with the navy. In *De Magnete*, Gilbert boasted of facts 'pointed out to me and confirmed by our most illustrious Neptune, Francis Drake',

Illustration 2: Portrait of Francis Drake (*c.* 1540–1596). Gilbert claimed to have consulted Drake, 'our most illustrious Neptune', in person about compass behaviour in the Southern Hemisphere. Drake, a buccaneer and explorer, led the first circumnavigation of the world after Magellan's historic expedition. As a vice-admiral, he helped to defeat the Spanish Armada in 1588. He embodied Elizabethans' pride in their new nautical supremacy.

and elsewhere he mentioned the 'most skilful Spanish, English and Belgian sea captains, with whom we have often conversed'.

Gilbert learned all he could from navigation experts. He himself probably never even went to sea, let alone navigated by compass, but during the 1580s and 1590s he built up a knowledge of geomagnetic phenomena far surpassing that of any other natural philosopher. Gilbert's most important colleagues here were two of the authors he mentioned, William Barlow and Edward Wright.

Like Gilbert himself, Barlow and Wright were educated gentlemen who had been drawn to maritime London. The Reverend Barlow's family was a veritable production line for bishops. But William diverted some of his energy to magnetic navigation and, although he initially hated the sea, published in 1597 *The Navigators Supply, containing many things of principal importance belonging to Navigation, and Use of Diverse Instruments framed chiefly for that purpose*. Wright was a Cambridge mathematician, who was asked to join the Earl of Cumberland's expedition to the Azores in 1596. On his return, considerably delayed by bad weather and botched navigation, he worked in London as a lecturer in navigation for the East India Company. In 1599, he published the fruits of his deliberations as *Certaine Errors in Navigation*, which remained the best English book on the subject for decades.

Wright and Barlow agreed that Gilbert had been working on magnetic philosophy and experiments for many years before 1600. Wright suggested that Gilbert had formed his basic ideas in the early 1580s. In the 1597 *Navigators Supply*, Barlow recorded that:

understanding by conference with a man of rare learn-
ing both in Phisicke, his owne profession, and in other
divers laudable knowledges, besides that, he many years
hath laboured in the consideration of the properties of
that Stone, and mindeth now out of hand for the com-
mon benefite to publish those his labours, I surceased
altogether from that purpose of mine, assuring me that
hee (if any other) will be able most exactly to handle
that Argument: for I found him excellently skilled, farre
beyond anything that I either knowe or imagined in that
matter.

The research on which Gilbert 'many years laboured' was
certainly time consuming and expensive. Not many
could have refuted the report of his contemporary, the
natural magician Giambattista della Porta, that a needle
rubbed on a diamond also pointed north, but Gilbert
'made the experiment ourselves with seventy-five diamonds in the presence of many witnesses'. His finances
were getting better by the year, as his practice flourished
and as he inherited some of his father's property.

The Glory Years

By the 1590s, Gilbert was living in luxury in Wingfield
House, a mansion close by St Paul's Cathedral that had
probably come from his stepmother. It was in the heart
of the city, close to booksellers and instrument makers,
with court bureaucrats for neighbours. Gilbert kept his
own little 'company' of lodgers there, and must have
installed a laboratory, cabinet of curiosities and library,
no doubt with a small household retinue to look after his

body and mind – and to do really sweaty experimental work such as smelting iron. Gilbert was a gentleman and courtier, after all.

His most famous lodger was John Chamberlain, a famous leisured writer of gossipy political letters. 'Dr Gilbert's house' was Chamberlain's London *pied-à-terre* from 1595 to 1601, and a couple of glimpses of Gilbert appear in his writings. These don't quite tally with the only other eyewitness evidence we have. Forty years after his death, his first biographer, Thomas Fuller, found a relative who said, presumably of Gilbert's mind, that he had had 'the clearness of Venice glass, without the brittleness thereof; soon ripe, and long lasting, in his perfections'. Fuller also heard that '"he was stoical, but not cynical"; which I understand reserved but not morose', and reported that 'his stature was tall, complexion cheerful; an happiness not ordinary in so hard a student and retired a person'.

These attributes are probably what Fuller expected from England's only world famous natural philosopher. Chamberlain hints at a man more like the acerbic writer of *De Magnete*: gregarious, amusing, opinionated in politics and religion, and certainly cynical in the modern sense. He was scathing about one courtier's attempts to get rich quick on an expedition to seize Spanish bullion: the booty, thought Gilbert, could be carried home on 'a well sadled rat'. And as scandal erupted in 1612, when a preacher denounced open Catholicism in James I's court, Chamberlain remembered that 'it is not good *irritare crabrones*, or to meddle with these pulpit-hornets, as our Doctor was wont to call them'. Unfortunately, no one is going to build a biography out of scraps like these.

We do know, though, that as the *annus mirabilis* of 1600 arrived, Gilbert was financially secure, with his city mansion, Tymperley's in Colchester, and nine other substantial properties with lands in East Anglia. His medical practice took him to court, to clients in noble entourages dotted around the city, to the docks and to the College, but it left him plenty of leisure. He had no wife, few family responsibilities and a circle of expert acquaintances with whom to refine his manuscript work of magnetic philosophy.

In 1600, the fellows of the College of Physicians elected him President. Then came the call from the

Illustration 3: Tymperley's.
Tymperley's was the Gilberds' Colchester home, which Gilbert inherited. Part of the fine Tudor town house survives and is now a clock museum. His London mansion, Wingfield House, burned down in the Great Fire of 1666. Dr Gilbert died a rich, propertied man.

Queen in February 1601. He was to move out of Wingfield House and into court. In between, he sent *De Magnete* to the press – a journey of a few hundred metres from his house to Peter Short's shop in Bread Street by St Paul's. Gilbert knew the book was explosive, and may have been waiting until his career was secure before he published his patronless philosophy.

Gilbert's brief life after *De Magnete* is equally undocumented. He attended Queen Elizabeth right up to her death in April 1603 where, according to Chamberlain (or rather his witness, who must have been Gilbert), she refused all physic. James I, who had a preference for chemical physicians, re-appointed him, which was not automatic. He collaborated with London's practical mathematicians upon the navigational applications of magnetic philosophy. He corresponded with William Barlow who, in order to establish his own credentials as a magnetic philosopher, published Gilbert's only surviving letter, which contained some praise of him. Gilbert told Barlow of his continuing magnetic research, a proposed addition to *De Magnete*, and his receipt of the letter from Giovanni Sagredo, who 'reporteth wonderfull liking of my booke'. This is the only evidence that Gilbert was aware before his death, during an epidemic of plague, on 30 November 1603, of the excitement that *De Magnete* was stimulating among continental philosophers.

CHAPTER 2

NAVIGATING THE MAGNETIC EARTH

According to Pliny the Elder's *Natural History*, magnetism was discovered when an ancient Greek shepherd called Magnes found that the iron nails of his boots and the tip of his crook were attracted by a rock. William Gilbert dismissed the legend, and so might we, but how much more than Magnes do most of us know about magnetism: we pick up two magnets and wonder at their attraction and repulsion. From the Greeks to the Renaissance, the magnet was likened to another strange wonder of natural attraction, amber resin. Amber attracted bits of straw when it was rubbed. It was a natural example of the electrostatic attraction that we can produce in man-made plastics and rubber.

What is Magnetism?

Today's scientists know that magnetism and electricity pervade all matter. Like the Greek atomists, who thought that lodestones and amber both emitted 'sticky' particles, they treat them as basically the same phenomenon, today called electromagnetism. It is a historical irony that Gilbert, who has been called the father of magnetic and electrical science, worked hard to establish how different they were, yet he would never have proved that the Earth was magnetic had he not done so.

Despite the sophistication of electromagnetic theory, still no one knows what magnetism really, *really* is. Physicists theorise that it is caused by a fundamental particle, the magnetic monopole, but no supercollider yet built has detected this magnetic equivalent of the electron. For the moment, our common knowledge that every magnet has a north and a south pole (that every magnet is a dipole) is true.

Since William Gilbert, we have known that the Earth is a magnet. Children's pictures show our planet as a ball with a big bar magnet stuck inside it. Gilbert already knew that was far too simple. The dipole model is a very rough approximation, but Gilbert knew from Elizabethan navigators and their knowledge of compasses that the truth was much more complicated – in fact, modern study of geomagnetism reveals the molten interior of the Earth to be the most complicated magnetic system we know.

On the much smaller scale of atoms, in our modern view, some elements are miniature dipoles. Among these, iron is special. Not only are its atoms dipoles, but also small numbers of atoms (about 10,000,000,000,000,000) cluster together in the same alignment to produce a microscopic magnetic island or 'domain' within a solid piece of the metal. Of course, most pieces of iron aren't magnets. This is because the domains themselves are aligned in many directions – so they cancel out or have zero 'magnetic moment'. But iron reveals a truly extraordinary property when it is placed in a strong magnetic field. The external field lines up the domains, and they remain aligned when the field is removed. Iron is not the only element to have this 'high susceptibility', but it is the most abundant. Moreover, humans have known how to smelt it from iron ores since *c.* 1500 BCE. Iron and its compounds were the only magnets known in Gilbert's time (if we discount the beliefs, which he rebutted, that diamonds made iron point north, and that many entities, including the organs of the human body, acted like lodestones. How else, argued most physicians except Gilbert, is the digestive system able to 'attract' nutrients into the body and to 'repel' waste and poisonous matter?).

But smelting iron doesn't make stuff you can use for permanent magnets. In pure, soft iron, the domains return one by one to random orientations. Elizabethan navigators knew they must remagnetise their compass needles on long voyages. To do so, they reached for the only permanent magnets they had – the natural magnets called lodestones (or loadstones).

Modern, strong permanent magnets are man-made. Super-susceptible alloys such as NIB (Neodymium–Iron–Boron) are subjected to powerful electromagnetic fields. Even the humble fridge magnet is 600 times stronger than the Earth. But ores rich in iron – oxides or sulphides such as magnetite and haematite – gave a plentiful enough supply in the past. Gilbert's era knew them all as 'loadstones', although Gilbert knew and described different kinds. Modern scientists think of the permanent magnetism in lodestone as an unusual form of a universal property of matter. For Gilbert, lodestone was a unique substance, 'a true offshoot of mother Earth' as he put it.

Magnets come in all shapes: discs and horseshoes are common today. Gilbert believed that the ideal shape was a sphere, because the Earth was spherical. In the seventeenth and eighteenth centuries, a spherical lodestone was the *pièce de résistance* of a gentleman's magnetic experiment set. Today they are museum pieces: most physicists have never used a spherical magnet. The bar magnet has become the laboratory favourite. Gilbert agreed that bar magnets had the strongest poles, but feminists, if not physicists, would appreciate his insistence that magnetic potency (which he likened to sexual attraction) was diffused through the whole body and not concentrated in the tips of phallus-shaped rods.

Conventionally, magnetic lines of force – the ones you supposedly saw in school when you dusted iron filings on to a piece of paper with a bar magnet underneath – always flow between the two poles of a magnet. When applied to the Earth, the convention results in an amusing paradox. Logically enough, the north pole of a magnet (when you are on planet Earth) is the north-seeking one. That usage was established by sailors long before Gilbert established that the Earth was a magnet too. The north pole of a compass or lodestone must, therefore, have the opposite polarity to the Earth's 'north' pole. Since Gilbert believed that the Earth was the origin of all magnetism, he insisted that the north-seeking pole was actually a south pole, attracted to the true North Pole of the Earth. But conventional usage has stuck and modern scientists, for whom magnetism is universal and not confined to the Earth, are happy with the opposite resolution. So the Earth's north geomagnetic pole is actually a south pole, and lines of geomagnetic flux are held to run from south to north around its sphere!

Today we know that the Earth's internal structure, with a molten iron core and a solid mantle, which varies in thickness and mineral composition, makes it a very complicated magnet indeed. But since 1400, Europeans have thanked God for the guidance of compasses, which seemed marvellously, if not miraculously, to point to the North Pole. The first navigators believed, or hoped, that magnetic needles pointed true north in some direct and simple manner. From the late fourteenth century, the study of magnetism was driven by the alluring potential of magnets as position finders, even solutions to the famous longitude problem. How wrong they were, and how much they and Gilbert had to discover to make sense of magnetic direction. Not until the eighteenth century was the allure rejected as that of a Siren enticing seamen to shipwreck.

Latitude, Longitude and Magnetic Meridians

The geographers' terms of pole, equator, parallels of latitude and meridians of longitude are largely mathematical inventions, projected on to the Earth's surface in a geographer's imagination or drawn on a globe. They provide a convenient and accurate grid reference system – although most readers will understand better that I wrote this book in Lancaster, just north of Manchester, England than at 54°N 3°E.

To the vast majority of mathematicians in Gilbert's time, who believed that the heavens and not the Earth rotated, the terms could have no Earthly reality. The sphere of stars spun daily on its giant axis, around the celestial poles. These were points near the Pole Star and the Southern Cross, where the daily orbits of the stars became vanishingly small. In the middle of the universe sat the stationary Earth. As the giant celestial axis cut through the Earth, it marked what we call the Earth's North and South Poles. But because the Earth was stationary, these points or poles had no geophysical significance. It was the heavenly spheres, or rather space itself, that had physical polar points, joined by an axis with the centre of the universe as its mid-point. Any natural motion could therefore be classified in terms of straight-line motion to or from the centre, or a circular motion around the celestial axis. This concept of space is very different to our Newtonian concept that space is the same in all directions. Gilbert rejected it too.

The equator, of course, was the great circle equidistant from both poles. It formed a celestial plane that cut through the centre of the Earth at 90° to the axis. Points on the equator are therefore said to have a co-latitude of 90° or, more usually, a latitude of 0°. Other parallels of latitude up to 90°N and 90°S indicate the angle made at the centre of the Earth/universe between that parallel and the equatorial plane. The easiest way to find the latitude of a northern location (apart from looking it up in an atlas!) is the navigators' time-honoured way of observing the 'angle of elevation' of the Pole Star – that is, how high it is in the sky above the horizon. This shows how parallels of latitude originated as projections on to the Earth's surface of celestial phenomena. It also shows how easy it is, in principle, to find your latitude. Easy in principle, but difficult to find at all, let alone accurately, on a storm-tossed boat in cloudy weather. No wonder at the excitement that met Gilbert's claim that you could find your latitude using just one new kind of magnetic needle.

Meridians of longitude are half circles running from pole to pole. (A complete circle of two meridians forms a great circle of longitude.) There is no simple way of finding your longitude from observations and calculations as there is for latitude. This is the origin of 'the longitude problem' that preoccupied many of the characters in this book. Consider places on the same parallel of latitude – say, Lisbon and New York. While it is still daytime in America, the Sun sets in Lisbon and the stars shine out from their usual elevations. Five hours later the same happens in New York at sunset, local time. The heavens look exactly the same, but at different absolute times, as they do at all the points between in the Atlantic.

It was well known in Gilbert's time that a solution to the longitude problem was to carry star tables and an accurate sea-going clock, set to the time of your home port. But, as Dava Sobel's *Longitude* explains so clearly, the technological problems of building such a clock were not solved until the mid-eighteenth century. The equivalence of all meridians also explains why nations disputed which meridian should be the prime one. Portugal and Spain warred over where the line should be drawn between their empires. The Greenwich meridian (and Greenwich Mean Time) represents the triumph of Victorian scientific imperialism.

This was the geographical grid of Renaissance mathematicians. Ours differs in that we agree with Gilbert, the follower of Nicolas Copernicus. It is the Earth, not the stars, that rotates daily. There is a terrestrial axis, and the

Earth's poles are a result of the Earth's own physics. But there is no celestial axis and no privileged polar points in space. So Gilbert argued that the Pole Star in Ursa Major, although it is useful for finding north, is not physically special or magnetic just because it is near 'the pole'. If the Earth's axis shifted, another star would take its place.

In their quest to find latitude and longitude, navigation theorists hoped that the compass would reveal the existence of magnetic meridians alongside geographical ones. If they existed, then comparing the two would give a quick positional fix. Gilbert and his followers rightly deplored the idea, but it was not a hopeless one.

In modern terms, the lines of magnetic flux flowing between the Earth's poles also form a kind of grid, which maps the three-dimensional geomagnetic field. They flow (remember) out of the south geomagnetic pole and into the north pole, both of which are deeply buried and tilted. Wherever a compass is put, it will try to align itself exactly with the lines of flux. This was the principle behind Gilbert's invention of the *versorium*. A versorium was a miniature iron needle with a (non-ferrous) point mounting that allowed the needle to rotate freely in all directions. The magnetic direction it ultimately assumes is a complex vector quantity; the needle does not behave like an obedient gundog, lying flat on the Earth's surface and pointing true north.

To begin with, because the Earth's dipole is tilted at about 11° to its geographical axis, the needle points away from true north. This 'magnetic declination' is defined as the angle between the geographical and magnetic meridian at any point on the Earth's surface. In Gilbert's time, it was commonly known as 'variation'. The precise angle is impossible to predict, because the Earth's magnetic field is not uniform.

Variation had been suspected for decades. Navigators became adept at checking the magnetic meridian of their compass against astronomical determinations of true north, and correcting for variation before they followed a course using compass bearings.

If the geomagnetic field were perfectly uniform (which it is not) and magnets therefore pointed directly to the geomagnetic pole (which they do not), then variation would also be regular. One can imagine a unique great circle of longitude that connects the two geographical and two geomagnetic poles. The two magnetic meridians that made up this circle would coincide exactly with the geographical meridians. For navigators they would be 'true meridians', where variation was zero. Elsewhere, however, the geographic and magnetic meridians would intersect at an angle. As one sailed into longitudes east (or west) of them, a westerly (or easterly) variation would build up. The exciting consequences are obvious. Magnetic variation would correlate with longitude. The compass would solve the longitude problem. As a bonus it would settle the question of which prime meridian cartographers should use! But, alas, the god of geomagnetism did not oblige, although, as we shall see, many were convinced that he had. The first to flesh out the theory was a Frenchman, Guillaume de Nautonnier. His interlocking grid of magnetic and geographical meridians is shown in Illustration 13.

Another problem was the vertical component of the magnetic direction. A compass needle dips down below the horizontal because the lines of flux themselves are rarely parallel to the Earth's surface. As we now see it, they dip into it on their curved route between the subterranean dipoles. This component is called 'magnetic inclination', or 'the dip of the needle'. Inclination upsets an otherwise perfectly balanced compass needle as soon the needle is magnetised. Compass makers had pragmatically rebalanced their needles for years before Gilbert's contemporary, Robert Norman, announced that it was a real phenomenon in 1581. Gilbert's revolutionary theory explained it, but he went much further. He proved that inclination increases gradually from 0° near the equator (where the latitude is also 0°) to 90° at the poles. He deduced that the equator and other parallels of latitude had a real, magnetic origin. The practical application of his results was equally revolutionary – his inclinometer was a magnetic way to find the latitude.

The empirical data upon which modern theorists of the magnetic Earth rely are incomparably better in quantity and quality than the data available to Gilbert. It has revealed the Earth to be a magnetic system of extraordinary complexity that is little understood. The modern consensus is that the Earth's rotation sets up currents in the molten core, and creates an electromagnetic dynamo. The currents are arranged more or less symmetrically around the Earth's axis of rotation. If that were the whole story, Gilbert would be partially vindicated in his conviction, much derided since the eighteenth century, that magnetic poles were necessary properties of a moving Earth. But dynamo theorists invoke a number of factors, some secure and others speculative, to explain why the Earth's field is most like a tilted dipole.

Some experts believe that the present dynamo theory will soon join a long list of historical curiosities. If it does, Gilbert's work will still have pride of place in it. It was his genius to strip away the complexities and announce, for the first time, that the Earth is a giant magnet. He united the communities of philosophers, astronomers and navigators behind him. But, as we shall see, this was only part of his project. His greater purpose was to overthrow Aristotelian theories of the cosmos. To see why, we need to describe the system of thought that Gilbert was taught in Cambridge, and against which he rebelled so effectively.

· CHAPTER 3 ·

THE LOST TERRESTRIAL WORLD OF ARISTOTLE

Aristotelian natural philosophy of the Earth is all but forgotten today. Gilbert had a grudging respect for his enemy Aristotle, who flourished in the middle of the fourth century BCE, but he could not forgive him for having stifled more ancient philosophical speculations that the Earth was moved by its own vital power. He focused his assault on Aristotle's theory of matter, which systematically excluded the possibility. Gilbert had studied it twice at university, once in his BA course in natural philosophy, and again in more detail at medical school. The great medical authority in Gilbert's time was still Galen, and Galen had adapted Aristotle's theory to the human body. Gilbert had had specialist training, but non-experts like playwright William Shakespeare (and his audiences) knew the basics too. Those basics made up much of the popular science of the Elizabethan world view.

Gilbert had no respect for the Aristotelian scholars who shored up this shoddy science. Over the centuries, they had produced so many defences of it that, as Gilbert remarked with no sense of political correctness, 'even weak old women, blind men and barbers find it obvious, simple and easy to understand'. What was it about Aristotelian theory of the elements that Gilbert objected to? Just about everything in the paragraphs that follow!

The Terrestrial and Celestial Worlds

Aristotelians divided the universe into two worlds, each made of different kinds of matter. They assumed that Earth was at the centre of the universe and, looking upwards or outwards from it, saw the boundary marked by the orbit of the Moon. Below the Moon was the sublunary, or terrestrial, world, which was compounded out of the famous four elements of Greek philosophy, earth, water, air and fire. Above the Moon was the supralunary, or celestial world. This was made of different, more perfect, heavenly stuff. Exactly what kind of stuff was a hot research topic in 1600. The traditional answer was a transparent, crystalline solid, a fifth element or quintessence.

Since heaven and Earth were materially different, they obeyed different laws of physics. Aristotle pointed out that our terrestrial world was a transient world of change. Consider a tree. It grows from a seedling into a large organism before dying and decaying. In every year of its life, leaves and fruit seasonally come into being and pass away. Physics had to explain how terrestrial matter underwent such changes. By contrast, the heavens appeared unchanging. The night sky looks to us just as it did to Aristotle and to Gilbert, which is why astrologers can still work with Aries, Gemini and the other ten zodiacal constellations of Greek astronomy.

There are changes in the heavens, of course. Viewed from Earth, the Sun and stars rise and set daily. Superimposed on this, we see the Moon move monthly through the constellations. Dwindling numbers of us have the common Renaissance skill of tracking other planets like Mars in their slower, more complicated progresses through the zodiac. But the point about these astronomical cycles, unlike the life cycle of a tree, is that the stars (and planets) seem to move in perpetual and predictable circles around the Earth. No Earthly motion could match this perfection. The trajectory of the most powerful cannon ball had a starting point in the gun barrel, and an end as gravity pulled it back down to the ground.

By Gilbert's time, Aristotle's two realms of physics had long been endorsed by Christian theologians, much to his disgust. There was a very good fit. The Earth that God created for us was marred by sin. The terrestrial world had become a location of moral as well as physical corruption, while the heavens remained so much closer to God and to unchanging perfection. Theologians sometimes described the Earth as *faeces mundi*, the universe's excrement. The Earth was clustered into a ball at the centre of the universe, they explained, because such a cosmological arrangement separated most efficiently the doodoo from the divine.

Terrestrial theology was buttressed by terrestrial physics, specifically with Aristotle's theory of gravity. Aristotle argued that two principles of motion controlled the motions of the four terrestrial elements. These were a pair of 'opposite qualities', gravity and levity. Any observer could tell that earth and water were the heavy elements that tended downwards. They possessed gravity. Air, and especially fire (think about a candle flame), moved upwards due to the force of levity.

Gravity and levity were the physical principles that moved stuff to its natural place. For earthy things this was the centre of the universe. According to Aristotle, a stone falls down because it is programmed by gravity to seek the centre, which it will do until it comes to rest in its natural place. In Aristotle's matter theory, the Earth's sphere cannot perform a Copernican orbit around the Sun, or move in any significant way. That would break the laws of nature. Gilbert was determined to disprove those laws.

Distinctions, Distinctions!

The Renaissance world was composed of contraries: heaven versus Earth, perfection versus corruption, up versus down. Dualities regulated all areas of life, dividing man from woman, rulers from ruled, the sacred from the profane. And matter theory formed the natural philosophical centre of this paradigm. Why are there four and only four elements? Aristotle explained. Elements are formed from the combination of basic matter and two more pairs of opposite qualities, heat and coldness, moistness and dryness. These create four permutations. Matter plus coldness and dryness makes earth. Water is cold and wet. Fire is hot and dry. Air is hot and wet. (Gilbert, who endured several fenland winters at Cambridge University, remarked that it took a Greek to insist that air was naturally hot.)

The four qualities were also the foundation of the Galenic medical theory that Gilbert learned. In the body they combined to form the four humours. Hot and wet blood was the analogue of air, for example. A healthy person's humoral qualities were balanced, or tempered. If someone was distempered by a fever, then letting hot blood, or eating 'cold' foods, were rational treatments for restoring the balance. Wherever Gilbert turned, the 'four element' theory stared back at him, explaining the world with a coherence that was understood even by the common barber-surgeons he policed.

Bear with the Aristotelians for a few more of their oppositions. You can forget them and look them up when they become important at the end of this book. Aristotelian physicists were concerned with all kinds of 'motions', or changes. Preoccupation with locomotion, change of place, is a feature of modern physics. Aristotelians had theories of locomotion too. It followed from their 'two-worlds' model that heavenly motion and terrestrial motion were different. Both were defined according to different privileged points in the cosmos.

Heavenly bodies moved only in circles around the centre or, to be precise, around the great axis and poles of the universe. The complex motions of planets showed that their spheres strove for, but did not quite attain perfection. It was good enough for philosophers, who left mathematicians to struggle with the deviations.

The centre of the universe also defined the two natural kinds of terrestrial motion. There was motion towards it, caused by gravity, and away from it, caused by levity. Both were straight-line, 'rectilinear', and therefore imperfect motions with beginnings and ends. Lovers of distinctions will appreciate that there were even two kinds of these – complete and incomplete. Elemental earth was heavy enough to complete its motion to the centre, while water, not as heavy as earth, could only rest on top of it. Gravity and levity were called the locomotive qualities. The other four qualities divided into two 'active' or 'alterative' ones (heat and moistness) and the 'passive' ones of coldness and dryness. Consider the change when the temperature of cold water is altered by dropping hot iron into it. For Aristotelians, the heat in the iron was the 'agent' of the change, while the cold water itself did nothing: it received heat as a 'patient'. Since 'earth' was cold and dry, that made it very passive indeed, which Gilbert did not like. Aristotelians looked for agents and patients in many kinds of change. When a lodestone attracted iron, some thought of the lodestone as actively altering the iron, by magnetising it, and then drawing the passive iron to itself. Gilbert didn't agree with this either.

Before Gilbert, natural philosophers had no general concept of magnets. What they discussed, when they discussed it at all, was *magnes*, 'the lodestone'. And even then they usually discussed the wonderful stone as an example of something else. Galen had used it to justify his theory that the 'natural faculties' of the human body, such as digestion and excretion, worked by attraction. Gilbert was appalled by his sloppy use of explanation by analogy.

It was the best example of a whole category of so-called 'occult qualities' – the opposite of manifest qualities. Renaissance Aristotelians recognised that the four primary qualities (six, if you include gravity and levity) could explain manifest but not occult properties. Manifest means obvious and occult means hidden, and the lodestone clearly draws out the difference. Imagine that you are holding a magnet in your hand for the very first time. What do your senses tell you about it? What qualities does it obviously possess? It is *manifestly* heavy, dry, cold to the touch, and dark perhaps – like many other minerals. Now, in your other hand, bring another magnet up close to it. Wow! That was magic! Did you feel the powerful attraction? Nothing you observed in the single magnet prepared you for it to reveal that property. Magnetic attraction is an occult quality.

The distinction between occult and manifest qualities is, mercifully, our last. The origin of occult qualities was hotly disputed in the sixteenth century. Real Aristotelian purists tried to reduce them to the primary qualities. Many, however, went along with Neoplatonists and natural magicians, who thought they must have a celestial cause, such as an astrological emanation from a star or planet. For magnets, the Pole Star was an obvious choice.

Magnets, Forms and Souls

On the Aristotelian's conceptual map, the lodestone was like Easter Island, a tiny location significant only for a few very unusual and baffling phenomena. A well-used and up-to-date textbook in Gilbert's time was *Physiologia Peripatetica Libri Sex* [*Six Books of Peripatetic [Aristotelian] Natural Philosophy*], by Johannes Magirus. He arrived at the lodestone in Book 5, chapter 2, paragraph 59. By this point, he had covered the principles of physics, the heavens, the elements, meteorology and hydrology (the sciences of air and water). Book 5 was on the Earth. Terrestrial things were divided into metals, stones, liquors, plants, animals and humans. Chapter 2 divides stones into common and precious.

The lodestone is one example, albeit peculiar, of a common stone. Magirus tells us that lodestones are generally reddish, and have a 'natural faculty' of attracting iron. Why they attract iron is, he admits, very difficult. Remember that iron, a metal, was a different kind of stuff. 'If someone said that it was because of their similar substances and a sympathy between them, he would not be far wrong.' He then rehearses some of the 'facts' that had become embedded in philosophy since medieval times. Lodestones need iron as a kind of nutrient. Garlic and diamonds disrupt the attraction. And that's about it. His students didn't learn about polarity or magnetic direction, let alone the compass. Paragraph over: on to gemstones.

Much earlier, Thomas Aquinas had tried harder. Saint Thomas, or the 'Angelic Doctor' to Counter-Reformation Catholics, was the brilliant medieval synthesiser of scholastic Aristotelianism and Christian theology, for which Gilbert could never forgive him.

To Aquinas, it was clear that elemental theory could not explain the lodestone's occult quality. Attraction needed something grander. As he put it:

Elemental forms, which are the lowest and closest to matter, have no way of working beyond the active and passive qualities, such as rareness and density and other such qualities, which are clearly material properties. Above these forms, however, are the forms of compound bodies which, as well as the aforesaid way of working, have another way of working appropriate to their kind, which comes from the heavenly bodies. Thus the loadstone does not attract iron by means of heat or cold or any such quality, but by means of its participation in the heavenly virtue.

Above compound forms came the souls of plants, and the animate forms of animals above them. Aquinas had placed the lodestone in between the inanimate and animate worlds.

In *De Magnete*, Gilbert quoted Aquinas at some length. If you found the last quotation difficult, try this one. But don't worry if you can't understand it. I think Gilbert intended us to dismiss it as academic verbiage.

> *A thing can be said to attract in another way, when it moves something to itself by altering it in some way, from which alteration it happens that the altered body is moved with respect to its place. In this way the loadstone is said to draw iron. For just as a generating body moves heavy and light things insofar as it gives to them the forms by which they are moved to a place, so does the loadstone give a certain quality to iron, by means of which it is moved to the loadstone.*

Gilbert agreed that Aquinas' explanation was not 'ill-conceived'; it acknowledged that element theory was not enough. He did not agree with the concept of the lodestone as the active alterer, or with Aquinas' likening of magnetism with gravity. Above all, Aquinas' experimental knowledge was woeful. 'This most learned man proceeds later briefly to corroborate [his explanation], citing incredible accounts of the loadstone and the power of garlic over loadstone.'

There was, however, one fundamental theoretical aspect of Aquinas' explanation that Gilbert did not overthrow – the Aristotelian concept of form. It needs some unpacking.

Aristotle adapted his tutor Plato's ideas into his concept of 'form and matter'. It provided Aristotelians with their general causal theory of change, which was very different to that of modern physics. Forms were the principles that 'informed' basic matter. Their differences explained why different substances, different beings, had their particular natures, properties and their potential to act as agents of change. Matter itself was entirely passive. That is why Aristotelians found atomism and the modern, mechanistic philosophy of Descartes so implausible. How could the mere motion and collision of particles of matter explain phenomena such as temperature or the circulation of blood?

As Aquinas made clear, there was a hierarchy of forms and the substances they explained. Elemental forms consisted only of the basic qualities (heat, moistness and so on) and their derivatives. They conferred an active 'virtue' upon the elements, but it was very limited. The forms of more complex but still inanimate substances such as stones were augmented by heavenly virtue.

Animate beings had even greater complexity and virtue. They had the ability to ingest and digestively alter food. Even plants had this faculty of nutrition. Animals could move in and respond to their environment. Humans (and angels) also possessed the faculty of reason. Aristotelians inferred from such observable properties that the form of animate beings included a life principle called soul (*'anima'* in Latin, from which we derive words like 'animal' and 'animation'). Their hierarchy of forms therefore continued with vegetative, sensitive and rational souls.

Natural philosophers empirically distinguished

animate beings from inanimate substances by their greater complexity and agency. For example, the animate form of a plant organised its matter into distinct parts with different properties, such as roots and leaves. Unlike minerals, plants had a top and a bottom. Gilbert agreed that the form of plants was animate. He cited evidence from the gardener's art of grafting. A gardener knew that a grafted stem had to be joined in the right direction or it would not take.

Animal souls gave beasts the faculty of self-motion. It organised their matter to provide them with suitable organs or instruments of motion, such as the contractile muscles of a worm or human legs. Animal movements were not random. Some avoided daylight and predators. Others moved in co-ordinated shoals or herds. Their sensitive souls gave them instinctual sense, if not reason. Aristotle had made reason the distinctive virtue of the human soul. Christian Aristotelians like Aquinas agreed, although they rebuked Aristotle for his dangerous error of denying that the human soul was immortal.

Human souls survived death, and angels had no material bodies. These Christian facts were proof that souls were immaterial. Aristotle had agreed, with Plato and others, that souls and other forms were immaterial principles. He just disagreed that forms could exist apart from the matter they informed. For everything except the immortal part of souls, Renaissance Aristotelians agreed. Gilbert himself cited a good, medical proof. When the human body was badly burned 'the soul's primary powers are not burned, though yet the burnt body remains without faculties'.

Like many other aspects of Aristotelianism, the

general ideas of plant, animal and human souls, and lower forms, were part of the common fabric of the Renaissance world view. Gilbert shared them. His radical act was to give the Earth a soul and an elevated place in the hierarchy of forms. Readers might divert themselves at this point by trying to enter Gilbert's mind and predicting the precise analogies that he drew in *De Magnete* between souls and magnetic force.

Whether the heavenly bodies had souls or 'intelligences' was more controversial. Influential Islamic Aristotelians had said so, and so did some Renaissance Neoplatonists. Putting souls in planets was one way to explain their apparent capacity to move themselves in regular, circular orbits which, according to some Platonists, added up to a co-ordinated 'harmony of the spheres'. Against this view, most Aristotelians had traditionally argued that the heavens were a giant machine, a nest of solid spheres driven by an external 'prime mover'. In Gilbert's time, the growing uncertainty about whether the heavens were solid, fluid or even a vacuum gave a renewed plausibility to the idea of planetary 'motor souls'. Johann Kepler, the Copernican astronomer, promoted the idea in his first book, *Mysterium Cosmographicum* [*The Cosmographical Mystery*]. He published it just before *De Magnete*, in 1597. It had no impact upon Gilbert, but Gilbert's magnetic soul had a profound influence upon him.

CHAPTER 4

PEREGRINUS AND THE DISCOVERY OF MAGNETIC POLES AND COMPASSES

At the same time as Aquinas was battling with the form of the lodestone, and battling to defend his theology, one person, Petrus Peregrinus (Peter the Pilgrim, aka Pierre de Maricourt), was on a real battlefield. He seems to have been a military engineer on crusade. Working independently, he produced an extraordinary '*Epistola de magnete*' ['Letter on the lodestone'], written on 8 August 1269 to his military friend Sygerus when he was at the seige of Lucera in Italy. It is all we know about him. Even more than *De Magnete*, this rare example of medieval experimentalism, and philosophical interest in the lodestone, came out of nowhere. Like *De Magnete*, the letter was a rare combination of natural philosophy and practical expertise. The potency of the combination, exemplified by Gilbert, ultimately transformed many sciences.

Peregrinus was educated enough to use the language of scholastic physics, but he scored over Aquinas through his familiarity with the lodestone's polarity. It is possible that he saw some of Europe's first compasses, used by the Italian sailors who ferried crusaders to the Levant. Peregrinus' work was a potential mini-revolution, but it sank largely without trace. Even Gilbert conceded that the thirteen short chapters made up 'a pretty erudite book considering the time'.

Peregrinus is important in Gilbert's story for three reasons. First, his was the first known account of a magnet's bipolarity, the first, in fact, to talk of magnetic poles. Second, he explicitly linked the lodestone to the cosmos. Where Aquinas had written generally of the lodestone's heavenly virtue, Peregrinus used his knowledge of its north- and south-seeking property to argue that this virtue flowed into magnets from the heavenly poles. Finally, his experimental design and apparatus provided the inspiration for Gilbert's own hypothesis and research.

Peregrinus described a series of experiments for identifying the lodestone's poles. He instructed Sygerus to have a lodestone turned into a sphere, 'using the technology for making crystals and other stones round'. He inspired Gilbert to revive spherical lodestones, and to play down the more obvious polarity of bar magnets. His cosmological reason, however, could not have been more different. North, South and the other points of the compass were not primarily properties of an Earth-bound magnet, but derived from the celestial poles and axis, and the heavenly spheres that swung around these privileged parts of the universe. And so he concluded that the lodestone acquired its polar properties from the celestial spheres. In fact, it imitated them, meaning that the best shape for experiments with lodestones was a sphere.

Having apparently complicated matters by using a spherical lodestone, Peregrinus explained how to find its poles experimentally.

An iron needle is put on the stone, and, following the length of the iron, a line is drawn dividing the stone in the middle. Then the needle is put in another place and

another line drawn ... You can do this in many places
and without doubt all the stone's lines will converge in
two points, just as all the meridians of the world's
spheres converge in the two opposite poles of the world.

Peregrinus then showed how to tell the lodestone's
north from its south pole. Unlike a well-mounted compass needle, a lump of magnetite doesn't just swing into
alignment. Peregrinus needed some form of frictionless
suspension. Ingeniously, he floated his lodestones on
little wooden boats. They obliged by aligning north–
south. Telling Sygerus to float two lodestones, he wrote:
'You will learn as a rule that the north part of one stone
attracts the south part of another stone, and the south
the north.' Like poles will flee. He confirmed his rule that
opposite parts attract by bisecting a stone into its north
and south halves. Division nevertheless produced two
whole magnets, each with two poles. The cut surfaces
became opposite poles, and so the two halves readily
reformed into the whole stone.

Peregrinus' experiments were prominent in *De Magnete*
– though Gilbert made no acknowledgement. But he
found the pilgrim's explanations as bad as those of the
scholastics. Peregrinus argued that whenever two stones
attracted each other, one was stronger. The stronger one
acted as the agent, and sought to draw and unite with the
weaker 'patient', until they became one. Finally, since
the lodestone was a likeness of the heavens, its north and
south poles received virtue from the celestial poles. Those
who thought it came from the Pole Star were wrong.

He finished with a speculation. Since the poles of a
spherical lodestone received virtue from the heavenly

north and south poles, then maybe 'the whole stone receives influence and virtue from the whole heavens'. In which case, since the whole heavens rotate every 24 hours, maybe a perfectly suspended lodestone does the same. He gave Sygerus instructions for this experiment in perpetual motion, but warned him to blame himself rather than a defect in nature if he couldn't reproduce this 'wonderful secret'. But Peregrinus does imply that he himself succeeded, and thereby inaugurated a trail of confident repetitions and failed replications of this 'fact'. Even Gilbert wanted to believe him.

Peregrinus' letter seems to have been saved from obscurity by his contemporary, Roger Bacon. Bacon was a friar at Oxford University with interests in alchemy, optics and, unusually, in experimental learning. A few scholars made manuscript copies for their libraries. When printing took off in the late fifteenth century, it wasn't exactly top of the must-read list. But by 1558, as Gilbert entered Cambridge, its time came. It had its fifteen minutes of fame when Jean Taisnier shamelessly republished it as his own work in 1572. Like Taisnier, Gilbert found Peregrinus' work still fresh and provoking. It showed how little time philosophers had devoted to magnetism in 300 years, and how few developments they had made. No wonder, then, that Gilbert preferred to learn from navigators.

The Early History of the Compass

A tidal wave of magnetic knowledge was released by the nautical compass. We do not know how the magnetic needle was discovered, but it was world-changing. Looking back in 1605, Francis Bacon observed that three inventions had transformed the ancient world into his modern one: printing, gunpowder and the magnetic compass. Bacon used them as emblems of progress. They were all unknown to the ancients, and all developed by practical men, not natural philosophers. They showed that Greek philosophy was not the summit of natural knowledge to be painstakingly reconquered by Renaissance scholars. Whole new intellectual worlds were there to be discovered. To discover them meant turning away from barren academic theory and harnessing the hands-on, experimental approach that had enabled navigators to discover a geographical New World unknown to Ptolemy. *De Magnete* was a product of the same revolutionary rejection of Renaissance values.

Bacon and Gilbert did not know that the Chinese invented the compass – and printing and gunpowder. The West exploited them differently and fully. All three were pressed into the service of imperial conquest, religious domination and commercial capitalism, while in China magnetic pointers were used to determine direction (often for the purposes of feng shui) and for geomantic fortune-telling. Magic was also a reason for Western interest. Not only did lodestones exemplify occult sympathies, but they also added to the court magician's repertoire of natural wonders and tricks. The West's mastery of ocean navigation made the difference.

Sailors probably started using magnetic needles in the mid-twelfth century. An early, full description was given around 1240 by Thomas of Cantimpré, a colleague of Aquinas. The practice seems to mix nautical know-how with magical technique.

When clouds prevent sailors from seeing Sun or star, they take a needle and press its point on the magnet stone. Then they transfix it through a piece of straw and place it in a basin of water. The stone is then moved round and round the basin faster and faster until the needle, which follows it, is whirling swiftly. At that point the stone is suddenly snatched away, and the needle points towards the Stella Maris. From that position it does not move.

Tradition accorded the honour of the invention to the seamen of Amalfi. An Italian origin is very plausible. Amalfi dominated Mediterranean trade (and the transport of crusaders like Peregrinus) until the rise of Genoa and Venice. All of these cities supported technical as well as scholastic learning.

The next significant development was made in fifteenth-century Flanders, with its trade and instrument-making centre in Antwerp. By this time, northern Europeans who lived in regions of higher variation knew that magnetic needles could not be made to point true north. The first instruments to embody the knowledge were portable sundials made in Germany. To set the sundial up it had to point north, and portable models included a miniature compass, with a mark on the base with which to align it. Many of these sundials have their marks about 10°E of true north. They are an unconscious record of the local variation at that time.

The Flemish compass makers had no concept of variation. They just wanted their compasses to leave their shops, and harbours, pointing north. They adapted the German solution. Old compasses let the needle swing freely above the card, the circular disc marked with the 32 points of the compass. The new Antwerp compasses had the card glued to the needle, but with the needle offset about 10°E. An ingenious solution, but one that unfortunately only worked in the maker's locality. As soon as a vessel carrying such a compass entered a region of different variation, the navigator set the wrong course.

Variation was about to be recognised as a serious problem. The Antwerp design spread to Genoa, where the needle was offset only 5°E. As international trade increased, in navigators and instruments as well as in slaves and spices, confusion reigned. When the expert Genoese navigator Cristoforo Colon (aka Columbus) defected first to Portugal and then to Spain, he took with him prized compasses made in several places, all with differently 'corrected' needles. The new generation of ocean navigators was up against the sharp end of the complexity of the magnetic Earth.

Chapter 5

Imperial Explorers and the Rise of Magnetic Navigation

In the two centuries preceding Gilbert's great work, the expanding empires of Portugal and Spain, later to be rivalled by France, England and the Dutch Republic, despatched sailors south, east and, eventually, ever further west. Navigation became a preoccupation of states and scholars more than ever before, both for charting new-found lands and trade routes, and for securing them as possessions. The big burst of exploration, culminating in the transatlantic voyage of Columbus and Magellan's epic circumnavigation of the world from 1519 to 1523, highlighted a host of problems related to the simple matter of how to locate oneself on the Earth's surface. One of the most vexing was the problem of magnetic variation.

Magnetic Variation, or Problems with the Equipment?

As one sails the Earth, variation changes slowly but appreciably. It is especially noticeable if one sails along a parallel of latitude. Even a relatively short voyage across the Atlantic can see the compass swing by 10°. This was noticeable to the most rough and ready navigator. And good navigators did 'run the latitude', because it was an easy course to check astronomically.

When Portuguese navigators began to use compasses routinely in the later fifteenth century, the needle was no more than a useful rough check upon direction, especially in cloudy weather. Navigators breathed sighs of relief when skies cleared and they could rely upon instrumental or skilled naked-eye observations of the Sun or Pole Star.

What mattered was that compasses pointed north. They knew that needles rarely pointed true north, but they thought their instruments were corrected. In Europe at this time, all variation was easterly. Portuguese sailors certainly knew about it and talked of their needles 'north-easting'. There is no evidence that they or their instrument makers worried about or tried to explain the phenomenon. The early mariner's explanation, that the needle respected the Pole Star (itself offset from true north by a few degrees), was not adequate. This was not because of Peregrinus' cosmological argument – the celestial pole was a worse explanation. It was because the Pole Star rotated diurnally around the celestial pole, a motion not reflected in the needle. For navigators it was a practical problem of compass manufacture with a practical solution: it was not yet dignified as a terrestrial phenomenon called 'variation'.

The magical, poorly understood nature of magnetism added to the 'mystery' of compass making. Experts debated many possible sources of error. Ignorant of inclination, they thought the needle might be inexpertly balanced; it might stick on its mounting; rust might build up to impede or corrupt the rotation; the kind of iron was important, for some held their magnetism longer than others, which mattered on a long voyage.

Above all, the lodestone used to magnetise the needle was surely critical. Gilbert showed that it wasn't as critical as they thought, but compass makers, navigation institutes and ships kept 'good' stones, and stroked needles in standardised, almost ritualistic ways well into the seventeenth century. With so many variables in the construction, and so many examples of faulty or weakened compasses, a navigator like Columbus almost expected his instruments to differ. That is why they often sailed with five or more compasses.

Although the Portuguese were the first experts in compass navigation, it was the Spanish voyages to America that created variation as a truly terrestrial problem. Since the early Portuguese expeditions sailed up and down the east coast of the Atlantic, magnetic variation remained within expected, easterly limits. But when Columbus first 'ran the latitude' right across the Atlantic to the Bahamas, his navigators were disturbed to find that all their compasses stopped 'north-easting' and started 'north-westing'. Compared with the crews' fears of never making landfall, this was an easy problem for Columbus' smooth talk – he reputedly explained it as a result of the Pole Star's motion.

When the Portuguese rounded the Cape of Good Hope into the Indian Ocean, they encountered problems as well. As ships swung around the Cape, so did their compass needles. Some speculated that the needles were affected by a magnetic mountain. Their problem was memorialised in the alternative Portuguese name, 'Cap de Agulhas' – the Cape of Compass Needles. Reliable reports from navigators accumulated of the needles' shifts in direction, especially in the busy waters around the Azores and further south near Cape Verde. Experts in the training schools of Seville and Lisbon began to take an interest.

Magnetic Meridians and the Problem of Longitude

As Portuguese experts collaborated to join up the dots of locations with zero variation, they found themselves looking at a meridian of longitude running through the Azores. Moreover, variation seemed to decrease in a linear relationship as one approached the meridian from either side. The idea of a true meridian of zero variation was put forward by Joao de Lisboa and Pedro Anes in 1508, and printed by de Lisboa in 1514 in his *Trafado da agulha de marear* [*Treatise on the Nautical Compass*]. He hypothesised that the true Azores meridian (25°W of Greenwich) was part of a great circle that also bisected the Pacific Ocean. In between, as one approached 90°E or W of the Azores, he predicted that variation would rise gradually to a maximum of 45°, and then decrease. The dream of a magnetic solution to the longitude had begun.

At the same time, politicians became interested in magnetic solutions. In 1493, Pope Alexander VI had settled territorial disputes between Spain and Portugal by dividing the world in half along a mid-Atlantic meridian of longitude. All new lands to the south and east were to be Portuguese, to the south and west Spanish. Disgruntled, Portugal managed to push the line westward in the Treaty of Tordesillas of 1494. That is why Brazilians speak Portuguese. Locating the Atlantic meridian posed the usual longitude problems, but they intensified when

Illustration 4: Treaty of Tordesillas, 1494.

At the 1494 Treaty of Tordesillas, Portugal succeeded in moving west by about 1,300 kilometres the great circle of longitude that Pope Alexander VI had used to divide new territories between it and Spain. The problem of finding longitude at sea made it hard to determine, especially in disputed Far Eastern waters. Compass experts joined the dispute in the hope of settling it using a meridian of zero magnetic variation.

Spanish and Portuguese fleets met in the Pacific near the prize of the spice islands. Where was the diametrically opposite meridian? One solution was to find and use prime magnetic meridians. It prompted the first international scientific conference of magnetic experts!

De Lisboa's theory of a regular distribution of variation around prime meridians became an entrenched, though contested, part of Portuguese and Spanish thinking. Magnetic longitude schemes of the Iberian kind were published or presented to many European courts throughout the sixteenth and into the early seventeenth centuries.

Schemers were encouraged by the proliferation of rewards for a practical solution to the longitude problem. Philip II of Spain promised a big sum of money in 1567, which Philip III renewed in 1598. The Dutch Estates General offered 5,000 guilders plus an annuity in 1600. The French and English courts had no formal state prizes, but that did not stop petitioners, who reasonably expected a fortune if their schemes were accepted. The influence of Gilbert and Wright, who both dismissed regular theories in *De Magnete*, guaranteed rejection in England for decades.

Despite, or because of, the very approximate values obtained by sixteenth-century navigators, it was already clear that theorists were selecting determinations that fitted their scheme. They were driven by unjustified convictions, and the lure of gold. Authors drew on growing records of values to adjust the locations of the prime meridians, rates of change of variation and (therefore) maximum values for variation as they sought the perfect fit.

It is tempting to see in such schemes the first realisations of a tilted dipole. If all the predicted magnetic meridians are followed to their points of convergence, these are not the North and South poles, but points on the prime meridian some degrees of latitude distant from the poles. But sixteenth-century exponents never mentioned offset poles, and they seem never to have calculated any locations. Nor did they speculate on what, if it was not the celestial poles, the needle respected. The absence of these components suggests that they were not, in fact, imagining a global theory of magnetism, but modelling local patterns and related longitude fixes. Indeed, they probably still had no generalised concept of variation. Even so, the schemes had some predictive power, which Iberian experts set out to test.

Magnetic longitudes promised a revolution in navigation, but they depended upon more accurate measurements. According to de Lisboa, the variation west of Lisbon decreased by only 1° in 100 leagues. The accumulation of different values for the same place, and of values that conflicted with predictions, increased the pressure for accurate, systematic surveys. Variation was exalted from a practical nuisance to a global concept. For the first time, manuals appeared describing how to measure it.

Scholars of mathematics were drafted in to assist navigators, a development that England caught up with in Gilbert's time. In Portugal, Pedro Nunez, the Jewish *converso* (convert to Christianity) and ex-customs officer in Goa, was made Royal Cosmographer in 1529 and given a new chair in mathematics. He invented better instruments, especially a 'shadow instrument' that permitted combined observation of magnetic direction and solar altitude, and he gave advanced training in their use. Professor Nunez was one of the first experts to publish in Latin, and his *De Arte et Ratione Navigatione* [*The Art and Method of Navigation*, 1546] brought knowledge and problems of magnetic navigation to a learned, European-wide audience that included Gilbert.

The government gave the task of the first survey of variation, and a splendid Nunez shadow instrument, to his student Joao de Castro, Chief Pilot to the Portuguese navy. De Castro left for a three-year voyage to the East Indies in 1538, during which he made forty-three observations of unprecedented skill. They threatened to blow accepted magnetic wisdom out of the water.

The concept of a true meridian through the Azores and Canaries was the first to fall. As he sailed south from Lisbon along it, he found various values up to 6°E. Sailing round Cap de Agulhas, he confirmed that variation swung rapidly from east to west, but there was no zero variation until he reached Natal. Worse was to come. Unsettled by the results, he checked his new-fangled shadow instrument against three of his pilots' trusted compasses. They all differed. De Castro's first thought was that each compass had been differently corrected by their makers, but when he opened them they were all set to the true meridian. Had their needles' magnetism weakened in different ways? He remagnetised them with the same lodestone – to no avail. His confusion climaxed when, while surveying the island of Chaul, he placed a compass on a boulder. It swung from north to south, and back again when he picked it up, despite the absence of detectable magnetism anywhere on the island.

De Castro's results were copied into the numerous, secret Portuguese '*roteiros*', or practical route manuals. When these inevitably fell into foreign hands, his work had international consequences. Authors published his results, and began to add their own. By Gilbert's time, scores of values were available – though many conflicted. This was not (just) because some were badly made, but because they were made in different years. Given that secular variation was regarded as literally unthinkable, there was no reason to date them!

The results also led to serious questioning of Iberian expertise and hopes for a magnetic solution to longitude. Adding to the confusion was the Sevillan examiner of navigators and Royal Cosmographer, Pedro de Medina. His *Arte de Navegar* [*Art of Navigation*], published in 1545, was one of the most widely translated manuals of the period, and was possibly used by Francis Drake. De Medina remained convinced that needles respected the celestial pole. Variation was an error, he insisted, resulting from the different kinds of iron, lodestone and construction used in compasses – not to mention sloppy practice.

De Medina needed only slightly to exaggerate the widespread unease about unknown and uncontrollable variables that attended the mysterious interaction between lodestone and needle, needle and world. Even so, by 1545 it was no longer credible, let alone safe, for navigators to ignore the real distribution of variation. Criticism rained in on him and his reputation suffered. English navigators preferred the superior *Arte de Navegar* of his successor, Martin Cortes, who 'corrected' de Medina's views on variation. But the opinion persisted that some, if not all, of the variation came from the apparatus. Even Robert Polter, the English Master of Trinity House, said so in his *Pathway to Perfect Sayling*, which appeared five years after *De Magnete*.

As Gilbert began his research in the 1580s, the direction and variation of the magnetic needle was becoming less and less understood as increasing amounts of data flooded in. Navigation experts could not even devise pragmatic rules for themselves. The out-of-touch and outdated theories of natural philosophers offered no assistance. Experts became outspoken about the inadequacy of natural philosophers, nowhere more so than in Elizabethan England.

Gilbert and Elizabeth's Navigators

When Gilbert left university, the indigenous art of navigation that inspired him had barely taken root. When he was born, only a handful of Englishmen knew how to traverse oceans. The Privy Council rectified the problem by stealing foreign expertise. In 1547 they poached Sebastian Cabot, the pilot major of Spain and also the son of Giovanni Cabota, the Genoese-born navigator who, in the 1490s, had masterminded voyages from Bristol to 'the New Found Land'.

Cabot hoped to set up systematic training along Spanish lines. But both he and his successor, Stephen Borough, who had also studied in Spain, were frustrated. Even in the late sixteenth century, the English government was suspicious of, and refused to patronise, expensive, Spanish-style central organisations. English exploration and navigational expertise grew out of private ventures, backed by coalitions of nobles and fluid networks of client seamen and advisors. Traditions built up – men whom Gilbert knew, like William Borough and Thomas Digges, had older relatives in the business. It wasn't efficient, but England avoided the institutionalised sclerosis that overtook Spanish navigation. *De Magnete* crowned a series of English innovations.

Until the 1570s, England was catching up, replacing foreign experts and expertise with its own. Stephen Borough oversaw practical training, including that of his brother William. He persuaded some members of the Muscovy Company, founded in 1553, to back Richard Eden, a graduate with an interest in exploration, in a programme of commercial translation and publication. He began with Cortes' *Art of Navigation* in 1561. Succeeding editions had to have novelty. Extraordinarily, the 1579 edition was bound with Taisnier's plagiarism of Peregrinus. Gilbert probably first encountered Peregrinus' experiments in it.

Dr John Dee, more renowned (certainly in his own opinion) as a natural philosopher and mystical alchemist, was forced to find work as a navigation tutor and consultant. He coached Thomas Digges and William Borough in the mathematical foundations of navigation. Early relations between scholarly and practical experts were sometimes strained. Digges really wanted patronage for his fine Copernican astronomy. He was appalled by the rough and ready reckoning of mariners. In turn, William Borough was appalled by Digges' and Dee's willingness to propose (but never publish) grand theories of variation and magnetic longitudes. Dee tired of practical mathematics, and emigrated to practise occult philosophy for the Holy Roman Emperor.

By 1574, William Bourne, an unschooled but experienced navigation teacher, had published his *Regiment for the Sea*, a deliberately simple but reliable manual that became the English standard. Elizabethan sailors put their home-grown expertise to use. A few privateers like Francis Drake dared to sail south into waters claimed, and patrolled, by Spain. Gilbert's collaborator Edward Wright gained his practical experience accompanying the Earl of Cumberland on a raid to the Azores in 1589.

Missions into Spanish waters were atypical, even after the defeat of the Armada. English efforts concentrated upon safer, uncontested Northern regions. The Muscovy Company traded with Russia – lucrative enough, but hardly a match for the East Indies' spice trade. England was cut off from the Indies, because Portugal defended the Cape of Good Hope and Spain the Straits of Magellan. Thus began the heroic but fruitless search for a north-east passage around Russia and, with Martin Frobisher's voyage of 1576, a north-west one. Walter Raleigh's exploration and colonisation of Virginia (named after his virgin queen) was initially unsuccessful, but led to England's American possessions.

These enterprises meant that English navigators became experts in high latitude sailing. Magnetic variation generally increases with latitude, in part because the angle subtended by the geographical and magnetic poles gets bigger. Near the north magnetic pole it also alters rapidly over short distances. Unlike their Iberian competitors, English sailors were forced to confront the problem of variation in extreme form. Their response was rigorous empiricism and vigilant scepticism of simplistic theories.

Thus it was an Englishman, William Borough, who in 1581 first published a tract dedicated to compass variation and its measurement. *A Discours of the Variation* typified the attitude of almost all English experts. Magnetic variation was real; a real problem for navigators, but a problem with no natural philosophical solution, only a pragmatic one of careful observation and correction. He gave a fully worked example for London – the average value was 11° 16'E. Before him, William Bourne feared that more records would confirm 'the opinion that the compass doth varie by no proportion'. George Best, who chronicled the quest for the north-west passage, was slightly less pessimistic: 'Captaine Martine Frobisher diligently observed the variation of the needle. And suche observations of skilful Pylots, is the onlye way to bring it in rule, for it passeth the reach of naturall Philosophy.'

Members of London's navigation community began to lose their feeling of inferiority to natural philosophers. They gained confidence that their practical, problem-solving approach and their trust in instruments and mathematics could elucidate some of the causes of compass behaviour. They were encouraged by Gilbert and other learned professionals who had rejected the *vita contemplativa* of Oxford and Cambridge in favour of London's dynamism.

Robert Norman and Simon Stevin

Bound with Borough's *Discours*, and dedicated to him, was a remarkable tract that made a big impact on Gilbert. *The Newe Attractive* was written by Robert Norman, an instrument maker who styled himself as an 'unlearned mechanician'. That did not stop him dismissing past writings on magnetism as conjectures that were refuted by 'experience, and thereby reason's finger', and criticising scholars who promised much but achieved nothing. He was especially affronted by learned mathematicians, who had told him that mariners should stick to their art and not meddle in higher things.

Norman expressed the English sentiment that 'the longer a man seeketh [causes of the lodestone's properties] the more he shall mervaill, and yet never the nearer his purpose'. He also rejected regular theories of variation. But he had discovered a remarkable 'experimented truth'. It was the natural dip of the needle, which today we know as magnetic inclination.

Norman was one of Elizabethan London's most respected compass makers. Frustrated by having to rebalance his excellent magnetised needles, he decided to find the cause. Perfectly willing to philosophise, he rejected the suggestion of 'unreasoned people' that lodestones imparted a weighty influence or spirit to the north end; he insisted that magnetic virtue was weightless. He mounted needles on a vertical pivot, and found that the north end always inclined some 71.5° below the horizon. He had made the first inclinometer.

His explanation took him further into the natural philosophers' territory. His inclined needle showed that magnets did not orient with some point on the Earth's surface, like a magnetic mountain. Either the south end was attracted upwards, in which case the cause was a celestial virtue, or the north end was pulled down by something inside the Earth itself. Showing a good command of celestial explanations of magnetism, and their inadequacies, he chose his new terrestrial cause.

He named the internal cause the 'point respective'. He proved that the needle was not actually drawn by a 'point attractive' with an experiment so elegant that Gilbert gave it a prominent place in *De Magnete*. Norman stuck a magnetised iron wire through a cork. Then he pared the cork down until the ensemble had the exact buoyancy of water. He immersed the wire in the middle of a glass of water. The wire dipped in the direction he expected, but it was not attracted in that direction down to the Earth.

Norman was not sure where in the Earth his point respective was. He invited people in other parts of the world to measure inclination. He was confident that the results would produce lines intersecting at a single point. Needless to say, an obscure book in English by a mere 'mechanician' did not galvanise the international community to make and send any observations. Gilbert was the first person to develop Norman's ideas, and he was able to complete Norman's programme of worldwide observations without leaving his laboratory!

Norman's hypothesis that magnets respected his point had several failings that Gilbert would expose. Nevertheless, it was the first novel explanation since the Middle Ages, and the first to suggest a magnetic cause inside the Earth. Norman had also announced the first new magnetic phenomenon since variation. All from a London instrument maker. The centre of magnetic innovation was now London, not Seville, Lisbon or Dieppe, and Gilbert was part of it.

We cannot close this survey of magnetic navigation

before Gilbert without mentioning the Dutch. As English navigation flourished in the 1590s, the Protestant Dutch Republic emerged as Europe's fifth maritime power. It had recently fought for its independence from the Spanish Empire with English help. When England defeated Philip II's avenging Armada, Dutch mariners progressed from excellent pilots in Northern Europe to navigators of the south seas. They acquired first Portuguese, then Spanish expertise. They came across the same seductive magnetic longitude schemes that the English had rejected. When Petrus Plancius, the Calvinist preacher, cartographer and naval advisor sought to patent his own version, the Estates General set up a committee.

One member was the practical mathematician Simon Stevin, who turned his report into a pamphlet, *De Havenvinding*, published in 1599. Gilbert's collaborator, Edward Wright, had Dutch contacts, and realised its value. His translation, *The Havenfinding Art*, appeared in the same year. It appealed to the English because Stevin had tempered Iberian confidence in the regularity of variation with Northern European suspicions that there were no simple laws. Regular magnetic longitudes were out, but, he proposed cautiously, there might be irregular but fixed patterns to variation.

This was his idea. Good tables should be compiled of variation data at intervals along the major latitudes that ships ran. To get the scheme started, he published Plancius' collection. As long as navigators knew their latitude (which was easier), they could then compare a sequence of onboard observations with the tables. That would at least tell them whether the longitude of their destination lay to the east or west, and it might even give an indication of how close it was.

Wright and the English welcomed Stevin's scheme. Its only weakness was its thoroughgoing empiricism. Stevin was the first to raise an uncomfortable question. Did variation stay the same in all places at all times? His scheme depended on it. That it might not was something 'reason will not suffer us to think'. No one ever had thought so, and the discovery of secular variation in 1634 was a shock. But how could one be sure in 1599? No observation programme could hope to detect a local shift that might send a whole fleet to destruction. The promising new art of magnetic havenfinding needed the guarantee of a causal explanation. By 1599, Wright knew that Gilbert's magnetic philosophy could supply it. He helped Gilbert all he could to prepare *De Magnete*. He introduced him to works of navigation, gutted their data, advised on practical compass problems and much more besides.

Aided by the work of Borough, Norman, Stevin and Wright and their continental predecessors, Gilbert knew more about the magnetic Earth than any other natural philosopher. When he combined this knowledge with his programme of experiments and his literature review of other natural philosophers, *De Magnete* was complete.

Classical philosophers, of course, had not known about magnetic direction, let alone variation or inclination. His review of recent authors on these subjects did not take him long. There was little impressive work to find.

· CHAPTER 6 ·

'THE COMMON HERD
OF PHILOSOPHISERS'

Gilbert knew that his magnetic philosophy was revolutionary. Revolutionaries condemn the lackeys of the old order for their faults, and Gilbert was no exception. He belittled natural philosophers of the Renaissance as a 'common herd', obediently following the paths of tradition. Renaissance authors enjoyed a robust exchange of rhetoric, but Gilbert went beyond the limits. Only practical 'men of experience' escaped his vitriolic and far from stoic pen.

The worst sheep in his herd were Aristotelian and Galenic professors. The universities were full of 'philosophasters' and 'sciolists', academics who had 'sworn an oath to their master' and lost the capacity for critical thought. The ex-Cambridge fellow wrote that: '[T]he doctrine "he himself said, Aristotle said, Galen says" has entered all the schools of good learning. The student one would call educated has disappeared.'

There were others who made exaggerated criticisms of conservative universities, but Gilbert criticised them too. In Gilbert's lifetime, Girolamo Cardano was the most widely read anti-scholastic natural philosopher. Gilbert considered that his 'ponderous volumes contain nought that is worthy of a philosopher'. Cardano's equally influential opponent, Julius Scaliger, was so ignorant that 'gardeners laugh at him'. His 'abundant verbiage' had the penetration of a 'lead dagger'.

A serious rival to Gilbert was Francesco Patrizzi. He had published a *Nova de Universis Philosophia* [*New Philosophy of the Universe*] in 1591, which also tore into Aristotelian theory of matter. Gilbert recommended that the best answers to its 'paltry arguments' were 'wet sponges, not written responses'. It wasn't really a *new* philosophy but a book 'cobbled together from tatty ancient fragments'.

There were a few exceptions: Peregrinus, of course, and the radical Copernican Giordano Bruno, who was executed in Rome as *De Magnete* appeared. Gilbert also acknowledged his debt to Giambattista della Porta, whose *Magia Naturalis* [*Natural Magic*] of 1558 included a substantial chapter on the lodestone. Gilbert called it 'a very storehouse and repertory of magnetic wonders [that have] given occasion for further researches'. *De Magnete* proved that. But Gilbert found della Porta ignorant. The magician was given to 'the maunderings of a babbling hag'. Gilbert thought that Patrizzi was 'full of his own opinion', but so was he!

Not surprisingly, then, when Gilbert reviewed the existing explanations of magnetic polarity, direction variation and inclination, he found little of value to report. A less jaundiced reviewer would have struggled to say much more, because philosophers had not noticed how the phenomena revealed by magnetic navigation challenged their world views. Gilbert dismissed 300 years of their work in a few lines.

Those great elementarian philosophers and all their

progeny down to our day ... were ever seeking the causes of things in the heavens, in the stars, the planets; in fire, air, water and in the bodies of compounds ... [F]or this reason, the common herd of philosophisers, in search of the causes of magnetic movements, called in causes remote and far away. Martin Cortes, who would be content with no cause whatever in the universal world, dreamt of an attractive magnetic point beyond the heavens, acting on iron. Petrus Peregrinus holds that direction has its rise at the celestial poles. Cardano was of the opinion that the rotation of iron is caused by the star in the tail of Ursa Major. The Frenchman Bessard thinks that the magnetic needle turns to the pole of the zodiac. Marsilio Ficino will have it that the loadstone follows its Arctic pole. ... Others have come down [to earth] *to rocks and I know not what 'magnetic mountains'!* [The variation theorist] *Livius Sanutus, to a certain magnetic meridian; Franciscus Maurolycus to a magnetic island; Scaliger to the heavens and to mountains; the Englishman Robert Norman to the 'respective point'.*

Gilbert's roll-call suggests that he was more familiar with the Latin works of natural philosophers and learned physicians than vernacular works by navigation experts. But Cortes and Norman were exceptional in their ambitions. Gilbert's survey was comprehensive.

The theories fall into four groups: celestial bodies or points, magnetic mountains, prime meridians and Norman's 'respective point'. Norman's was so novel that it was virtually unknown. The theory of true meridians was another sixteenth-century innovation, but we saw earlier that it was not a causal explanation. It was made into one in 1602, when a Frenchman twisted Gilbert's magnetic Earth into the first true tilted dipole.

Celestial virtues and magnetic mountains had a longer pedigree. Medieval mariners were probably the first to locate the origin of magnetic virtue in the heavens. They had always been guided by the Pole Star and they assumed that their compass was too. Peregrinus intended his sophisticated theory of the celestial sphere and poles to replace it.

Celestial Theories

A celestial virtue was light years away from Gilbert's magnetic Earth, but there was good reason to think that one could be transmitted to the sublunary world. Sunlight reached the Earth, as did the astrological virtues of other heavenly bodies. The specificity of magnetic virtue, which apparently affected only lodestone and iron, was more problematic, as it was for any theory. But sixteenth-century writers had a repertoire of Neo-platonic and natural magical concepts to deal with it.

Virtues of heavenly origin, sometimes thought of as emanations from planetary souls, were supposed to permeate the whole world. Rays of astrological influ-ence, or streams of invisible spirituous matter, suffused many substances. They gave substances their occult qualities. They set up hidden sympathies and antipathies between things, which it was the magician's task to discover by experiment. The sympathetic attraction of lodestone and iron was an obvious example, the anti-pathy between lodestones and garlic a more dubious one.

The Neoplatonic doctrine of correspondences, which was widely held in Gilbert's time, provided a theoretical foundation for sympathies. The Creator had replicated the properties of the cosmos-at-large, the macrocosm, in various little worlds or microcosms. Humans were microcosms; for example, their seven major organs were each ruled by one of the seven planets. The small, central sphere of the Earth likewise bore analogies with the heavens.

Peregrinus had developed a general analogy – every part of the Earth imitated every part of the heavens. Renaissance philosophers tended to prefer more specific correspondences, such as those linking the seven planets, organs and metals. Gilbert knew that the Florentine Platonist and physician Marsilio Ficino had pioneered a system of magical medicine that sought heavenly 'causes remote' for many illnesses. Ficino's theory differed from Peregrinus' by making the north celestial pole a unique emitter of magnetic virtue, which then flowed into all magnets.

Gilbert wanted to refute the theory of a remote heavenly cause almost as badly as Aristotle's theory of passive Earth. Both ignored the Earth's own magnetic powers. But seventeenth-century magicians like Robert Fludd continued to cite the 'correspondence' between the north poles of the heavens, Earth and lodestones as proof of their natural philosophy of sympathies. In fact, they seized on *De Magnete* as conclusive proof of it!

Ficino's choice of the north celestial pole suggests that he was ignorant of the lodestone's 'north-easting', although Antonio Pigafetta, the experienced Italian navigator who survived Magellan's circumnavigation, published the same view in 1525. De Medina, who denied variation, continued it. By plagiarising Peregrinus, Jean Taisnier (and his English translator) intensified interest during Gilbert's formative period.

The fact of variation led some natural philosophers to doubt the celestial pole theory. Because Renaissance magicians attributed specific virtues to many celestial bodies, it was easy to find a better candidate. The navigators' choice of the Pole Star, a couple of degrees away from the pole, was an obvious one. Gilbert was wrong about Cardano – he chose the Pole Star too.

An objection to any stellar origin was that stars seem to rotate daily around the celestial pole, a motion not reflected by the needle. That did not stop Jean Fernel, a French physician who discussed the lodestone in his *De Abditis Rerum Causis* [*Hidden Causes of Things*], from identifying a star in the tail of the Great Bear.

The navigation expert Cortes attempted to save the theory of celestial origins from Fernel's and Cardano's error. Sure that variation was real, constant and regular, he needed the magnetic virtue to emanate from a fixed point. But all points in the heavens except the poles rotated. And so, in his *Art of Navigation*, he located it beyond the cosmos. Celestial magnetism was a theory in trouble. Earthbound magnets looked more promising.

Terrestrial Mountains and Points

The concept of a magnetic mountain was literally put on the map by the great cartographer Gerard Mercator. Legends of powerful mountains, islands or rocks capable of attracting boats and even pulling the nails from their hulls grew out of classical stories. They were elaborated in Arab folk tales, such as *The Thousand and One Nights*, and discussed by medieval geographers and sailors. Italians suspected that the island of Elba disturbed their compasses. In the sixteenth century, the idea developed into a single landmass situated some distance from the North Pole. It controlled magnetic direction and created variation.

Mercator was inspired by the Swedish geographer Olaus Magnus to include a 'Magnetum Insula' north of Siberia on his globe of 1541. He refined its position, calculating the intersections of trusted magnetic meridians. His famous and much imitated map of 1569, drawn according to his new projection, included two possible locations. His favourite, on the supposed Atlantic prime meridian, was copied by French and Dutch cartographers.

Julius Scaliger showed how the mountain hypothesis could be combined with the enduring belief that magnetism was celestial. He agreed with Cardano that magnetic virtue flowed from the Pole Star, but insisted that a magnetic mountain acted as its terrestrial receptor.

When Robert Norman announced his discovery of inclination, he argued persuasively that it undermined both celestial and mountain theories. Gilbert believed that his 'respective point' still shared a fatal weakness with them. All three were Eurocentric, fixated with the asymmetric assumption that compasses worked by pointing north. Bipolar magnets could equally well be pointing south or, in Gilbert's view, aligning themselves with both poles. The 'common herd' could have proposed a star in the Southern Cross or a mountain in Antarctica to account for the lodestone's south-seeking pole or the symmetrical performance of compasses in the Southern Hemisphere, but they did not.

The most telling weakness of all explanations before Gilbert's was not theoretical but empirical. Whether the needle was controlled by a celestial point, a magnetic mountain or prime meridians, in every case variation ought to 'observe a rule' and to be distributed regularly over the Earth's surface. The evidence that it did not was building up.

Gilbert was probably the first natural philosopher in history to deny that the interaction between a compass needle and the source of magnetic virtue was, in principle, simple and direct. He did it because he had immersed himself in the English tradition of navigation. English experts knew the unruliness of variation.

He wrote off his predecessors' explanations because they were 'at odds with every-day experience'. That was certainly the everyday experience of English mariners. English expertise took the best of 100 years of Iberian exploration, and added the innovations of men like Norman, Borough and Wright. Empirical knowledge of complex magnetic phenomena had increased so rapidly that theory had been left behind. The English feared that it 'passeth the reach of natural philosophy', at least of the common herd. Gilbert's revolution brought it spectacularly into reach. Magnetic philosophy showed how you could clutch the magnetic Earth in your hand.

PART II

DE MAGNETE: GILBERT'S NEW PHILOSOPHY OF THE EARTH

· CHAPTER 7 ·

WILLIAM GILBERT'S
BRAVE NEW WORLD

De Magnete is a strange landmark in the history of science. In his *History of the Inductive Sciences* of 1859, William Whewell judged that 'Gilbert's work contains all the fundamental facts of the science [of magnetism], so fully examined, indeed, that even at this day we have little to add to them'. A generous view, but *De Magnete* is stuffed full of detailed, even trivial, facts. They explain its appeal to Victorians like Whewell, and many today who think that scientists use an inductive method. Start from careful observations, extend them with numerous experiments and only then form a general conclusion. Gilbert's project has been described as 'finding out all there is to know about magnets'. The project makes sense if the fundamental nature of magnetic force has been established. Gilbert did establish it, but no Renaissance natural philosopher in his right mind would have started a collection of lodestones and thought: 'Let's spend eighteen years studying this occult property; it might be more important than we thought.'

Why did Gilbert research and publish 'on the load-stone'? Was it to ferret out ferrous facts? Of course not. Was the lodestone Gilbert's inspired choice to illustrate the general power of experimental method? *De Magnete* certainly promoted experimentation, but why *on the lodestone*? Was he driven by the problems of navigation to bend his abilities to the service of Elizabeth I? That is a better motive, but a subsidiary one. Was he, as Francis Bacon suggested, so obsessed with his ideas that he became a nerdy magnetism bore? We all know specialists like that. In fact, Gilbert was motivated by truly cosmic interests.

Gilbert did not write just the one book, *De Magnete*, but three. The other two books mention magnetism occasionally, but all three add up to a more grandiose agenda – the replacement of Aristotelian natural philosophy with Gilbert's visionary system of the world. Gilbert's theory of the magnetic Earth was the cornerstone of his new cosmology. Now that is a motive for spending an Elizabethan fortune on magnetic experiments.

Gilbert's 'other books' appeared under the title *De Mundo Nostro Sublunari* [*A New Philosophy of our Sublunary World*], commonly known as *De Mundo*. He never published or even completed them, but his half-brother William cashed in on his posthumous success by presenting the manuscripts to James I's science-loving son, Prince Henry. Bacon read them: a copy found with his own manuscripts was printed in 1651.

Since Gilbert did not authorise his brother's deeds from beyond the grave, we can't be sure how much editing William junior did, when they were written, or whether the Gilbert of *De Magnete* stood by them. It's pretty certain that the book called *Nova Meteorologia contra Aristotelem* [*A New Meteorology against Aristotle*] was early work, which he probably left unfinished around 1583. He may have been dissatisfied with it, so we will draw only two safe conclusions. Gilbert had turned away from Aristotelian matter theory long before 1600. At the same time, he already believed that the Earth and Moon were magnetic spheres, that the Earth's magnetic virtue made it rotate and that of the Moon caused the tides.

The second book, called *Physiologia Nova Contra Aristotelem* [*A New Natural Philosophy against Aristotle*], is a different ball game. It is Gilbert's stab at a complete and revolutionary natural philosophy of the Earth and other heavenly orbs. He worked on it from the 1590s right up to his death, during the period when he was compiling and revising *De Magnete*. In *De Magnete*, Gilbert mentioned topics, heat and types of attraction that he would investigate 'elsewhere'. They are developed in *Physiologia Nova*.

Physiologia Nova refers back to *De Magnete*, although it is not about magnetism, and contains no new magnetic experiments. Few people have read both books. *Physiologia Nova* seemed either old hat or outdated when it appeared in 1651. Lacking the status of 'classic scientific text', no translation has been published. To a reader of both works, *De Magnete* can seem like a technical appendix to *Physiologia Nova*. *Physiologia Nova* is a whole world view built on the assumption that the Earth is magnetic. *De Magnete* 'shows the working', with factual support and navigational applications. The conclusions, subplots, philosophical asides and unresolved mysteries of *De Magnete* all come together in *Physiologia Nova*. So let's look at it first.

Gilbert's Heavens

Gilbert's was an infinite universe. The idea was not new, but had only been seriously proposed in 1584 by

Giordano Bruno, the radical philosopher who was executed in Rome just as *De Magnete* appeared. Copernican astronomy had made the idea possible. If the Earth revolves and the stars stand still, there is no need for them to be confined in Ptolemy's solid sphere.

Matching Bruno's radicalism again, Gilbert declared that interstellar and interplanetary space was a vacuum. That abolished not just the solid stellar sphere that he denied, but the entire mechanism of spheres that conventionally moved the planets. He was impressed by recent astronomical studies of comets. Comets were assumed to be emissions of subterranean vapours that burned up in the atmosphere, but Tycho Brahe and Thomas Digges had observed one that passed through the interplanetary regions, unhindered by solid matter. We might gasp at Gilbert's modernity, but he had no evidence for the vacuity or infinity of the cosmos. Unlike the magnetic Earth, they were radical, inspired, plausible and consistent – and lucky – guesses. History treated many of his other speculations less kindly.

So what held the stars in their places or moved the planets in their regular orbits? Gilbert was sure that each heavenly body had a specific internal motive power. As *De Magnete* proved, the Earth's was magnetic. Gilbert described the Earth's magnetism as an indwelling virtue, a soul or an 'internal life'. It was an immaterial force that held the Earth's sphere together, propagated into empty space, and rotated it on its axis. The Moon's virtue was similar: this 'companion of the Earth' acted on it to cause tides. Gilbert was so interested in the Moon that *Physiologia Nova* contained the first lunar map.

From Earth and Moon, he generalised to all the

planets. Their virtues were analogous to magnetism, but specific to them. The Sun's was literally visible in its light, which Gilbert conceived in traditional manner as an immaterial radiation. A planet like Jupiter would have its own 'Jovial' virtue, controlling the Jovian sphere, but also extending its power into space.

Gilbert was sure that the Sun's virtue dominated and governed the planetary system. It acted to stimulate or 'incite' all the other planetary virtues to move their planetary bodies around it. The idea that the Sun's luminous power was the most important was not new: Aristotelians acknowledged that sunlight affected life on Earth far more profoundly than any other planetary influence. But Copernicus had given this as a reason for the Sun's centrality. Gilbert's achievement was to develop it into a very plausible dynamics of the solar system. The plausibility came from Gilbert's experimental study of magnetic force. Even in *Physiologia Nova*, Gilbert avoided any clear statement that the Earth orbited the Sun, but he was concealing his position. (His cleverest evasion was to observe that an infinite universe has no centre at all!) His dynamic theory, his repetition of Copernicus' own arguments and the testimony of his colleagues give him away.

Gilbert's Heavenly Earth

And so we come 'down to Earth'. Gilbert reserved his greatest scorn for the traditional division of the heavenly world from the terrestrial. What pleased him most about magnetic philosophy was its proof that the Earth possessed the same noble, vital, self-moving properties that Aristotelianism reserved for the perfect heavens. Gilbert's Earth was not a lifeless, heavy, sluggish, passive ball of *'faeces mundi'* condemned to rest at the centre of the universe. Using Copernicus' Neoplatonic language, he described how the Earth played its part in the harmonious concert of heavenly motions.

As in *De Magnete*, Gilbert targeted his main assaults on Aristotelian theory of sublunary matter, especially elemental earth. Gilbert organised Book I of *Physiologia Nova* around the textbook topics of Aristotelian philosophy: what are elements? What is fire, air, water? What are the prime qualities of hot, cold, gravity and levity? And so on. His answers ranged from 'They're not what you think' through 'They don't exist' to 'The concepts are laughable'. Galenic medicine also came under such heavy fire that Gilbert's relations with the College of Physicians would have been strained had he published. Aristotle, Galen and the common herd dispatched, Gilbert used the second and last book to develop his own theory of matter.

To be honest, Gilbert's new theory of the elements was weak. In the 'sublunary world', there was really only one kind of matter, only one element. This was the pure, homogeneous magnetic substance of which all but the Earth's surface and atmosphere was made. Gilbert gave his pure earth a natural moisture, which it needed to bind together. All moist and watery substances therefore derived from earth.

With so few kinds of stuff, Gilbert's 'moisture' had to do a lot of things. When it was separated from true earth it formed 'effluvia'. The most obvious was water or rather (since he rejected Aristotle's element of water) various kinds of 'aqueous humours'. They formed the sweet, potable waters of rivers and salty seawater. 'Oily humours' formed minerals.

Aristotle had explained the variety of substances as different combinations of the four qualities. Gilbert only allowed substances to vary in their heat or moisture. Since moisture was part of pure earth, heat was his only quality. Heat from the Sun or the interior of the Earth expanded watery effluvia into the constituents of the atmosphere – there was no such element as 'air'! Fire was not even a substance, but the visible sign of an effluvium that was undergoing extreme heat.

Gilbert's sketch of a theory of matter was an interesting alternative to Aristotle's or Patrizzi's. It was hardly the road to Mendeleev and the Periodic Table. Like all one-element theories, it was riddled with contradiction. Gilbert had not taken seriously the ancient Greeks' logical conclusion that the variety of the world could not spring from one substance. His terminology ('qualities', 'humours', 'true earth') remained Aristotelian. His evidence was anecdotal and derivative. It badly lacked the experimental rigour of *De Magnete*.

De Magnete did provide experimental support for one of his most extraordinary theories. The famous section on electricity concluded that an electric was a 'concretion' of aqueous humour. It attracted when a gentle heat released the humour as an effluvium. *Physiologia Nova* explained that atmospheric effluvia had the same attractive properties. They performed the task that we (and Aristotle) assign to the Earth's gravity. Things were drawn to the surface of the Earth like fluff to a rubbed balloon! A corollary of Gilbert's effluvium theory was that 'gravity' stopped acting at or just below the Earth's surface. One of the few new 'experiments' in *Physiologia Nova* corroborated it. Gilbert recounted how a horse-rider of his acquaintance had fallen down a deep well and emerged unharmed.

Even with such pitfalls in reasoning, *Physiologia Nova* is not a bad, uninteresting book. It just isn't *De Magnete*. It is no worse than the contemporary publications of Giordano Bruno or Francesco Patrizzi. They too were radicals who imagined a new universe, and they too found the task difficult. In short, *Physiologia Nova* belonged to an emerging genre of post-Aristotelian 'new natural philosophy'. Its author, Gilbert, fitted the established role and identity of a natural philosopher with grand ambitions to break the mould of Renaissance cosmology.

But Bruno, Patrizzi and the others did not write anything like *De Magnete*. They did not found new sciences. They did not effect a revolution. This turns the historical problem of *De Magnete* on its head. Readers of *Physiologia Nova* have tended to ask what possessed the pioneer of sound experimental, magnetic and electrical science to fall back into unfounded Renaissance specu-lation? They wonder why he tacked a dodgy book of animistic Copernicanism to the end of his brilliant proof of geomagnetism. They assume that the 'modern science' we find in *De Magnete* was self-evidently superior. They can't understand how Gilbert went on to write *Physiologia Nova*.

I prefer to think of *De Magnete* as the problem book. As a self-proclaimed treatise of natural philosophy, its focus on lodestones and iron was weird. Its consistent reliance upon experiments was revolutionary. Its command of technical expertise was unprecedented. Its integration of 'pure' natural philosophy and 'applied' navigation was pathbreaking. How, then, did Gilbert produce *De Magnete*?

CHAPTER 8

THE HISTORICAL ROAD
TO *DE MAGNETE*

The road to *De Magnete* probably began in the 1570s. By then, Gilbert had left Cambridge and become dissatisfied with Aristotelian and Galenic matter theory. His independent research began around 1569, with meteorological observations of winds and comets. They encouraged him to question Aristotle's application of the theory of hot, cold, wet and dry qualities, at least to the weather. In the 1570s and early 1580s, they grew into his research project of an anti-Aristotelian 'new meteorology'. Traditional meteorological topics such as winds, seas, rivers and tides, and Aristotle's effluvial exhalations from the Earth, led him to think about the nature of the Earth itself. There was a general trend among late Renaissance nature philosophers to grant the Earth a greater vital activity and nobility than did Aristotelians, and Gilbert became part of it.

By now Gilbert was practising medicine in London. Then as now, doctors' diagnoses often met with scepticism. The French playwright and actor Molière memorably satirised the common perception that the physician's repertoire of specific virtues and qualities were mumbo jumbo. Gilbert agreed that his profession suffered not just from a poor, rote-like use of Galenic theory, but that much of it was wrong. Galenism stated that the body's faculties, and many specific drugs, acted like lodestones,

Dolit Gul: Gilbertus. fr: Stulwood proprys manibus

GVILIELMI GIL

BERTI COLCESTREN-
SIS, MEDICI LONDI-
NENSIS,

DE MAGNETE, MAGNETI-
CISQVE CORPORIBVS, ET DE MAG-
no magnete tellure; Phyſiologia noũa,
plurimis & argumentis, & expe-
rimentis demonſtrata.

LONDINI
EXCVDEBAT Petrvs Short ANNO
MDC.

Illustration 5: Title page of the first edition of *De Magnete* (London, 1600).

Gilbert's classic work was printed by his near neighbour Peter Short. It was called *A new natural philosophy of the lodestone, magnetic bodies, and the great lodestone the Earth, demonstrated with many reasons and experiments, by William Gilbert of Colchester, a physician of London.*

attracting goodness or repelling toxic waste. Gilbert rejected this analogy of magnetic attraction, and it must have at least prepared him for his re-examination of it.

In the early 1580s, Dr Gilbert was moving in high London and court circles, which were full of humanist secretaries, fellow professionals and mathematicians, such as Thomas Digges, who did military and naval work for their patrons. It was more likely with these men than with Cambridge mathematicians of the 1560s that Gilbert encountered support for Copernicanism. He embraced it, or at least the moderate 'semi-Copernican' thesis that a central Earth rotated on its axis. The Copernicans' anti-Aristotelian positions that the Earth was moving and noble like the planets fitted his reasons for rejecting Aristotle's element theory. He was mathematical enough to understand the greater economy and harmony of the system that Copernicus had outlined in his first book, even if his detailed grasp was limited and conservative. But he understood clearly that no Copernican had given a physical explanation of how the Earth moved that answered the Aristotelians' objections.

Gilbert's encounter with the ideas of Petrus Peregrinus' letter on the lodestone must have been a defining moment. He may have read it in Richard Eden's English translation, appended to the 1579 edition of Martin Cortes' *Art of Navigation*. That appendix appealed to philosophically minded navigation experts like Robert Norman, who was the first to suggest a magnetic cause inside the Earth in 1581. He must have met such experts in the 1580s as fellow clients of the great nobles he healed, and with whom we know Gilbert worked in the 1590s.

Peregrinus provided Gilbert with his key concepts. The pilgrim had experimented on a spherical lodestone; he had treated it as a likeness, though crucially a likeness of the heavenly sphere not of the Earth; he had given magnetism a cosmic significance; and he had stated that the spherical lodestone rotated diurnally in imitation of the heavens. In one sense, all Gilbert had to do was to transfer the analogy from the heavenly sphere to the Earth itself – and that is, I think, exactly what he did in the early 1580s.

Such an inversion from heaven to Earth would have been almost literally unthinkable to anyone before him. It attributed celestial virtue to the Earth. We saw how Gilbert's era continued to make heavenly bodies the origin of virtues such as magnetism. Renaissance Neoplatonism, the source of many criticisms of Aristotelianism, had even increased the tendency. Gilbert, however, outdid other Renaissance naturalists in his conviction that the Earth was the equal of the planets. It obviously called into question Aristotelian theory of passive elemental Earth. The inversion also implied that the Earth, not the heavens, rotated diurnally. It could only appeal to a Copernican (or semi-Copernican) – and Gilbert was one of the first. Finally, it made the whole Earth possess magnetic properties. Only a natural philosopher who was familiar with the blossoming art of magnetic navigation would have thought of that.

If a latest date of 1583 for Gilbert's *New Meteorology* is reliable, it must have been around then that Gilbert had his big idea of the magnetic Earth. There is one brief allusion to it, in his closing account of tides:

There are two principal causes [of tides, he wrote]; *the Moon and the diurnal revolution. The Moon does not impel the seas by rays or light. How then? To be sure, it acts by a mutual interaction of* [its and the Earth's] *bodies, and (to explain it by a similitude) Magnetic attraction.*

Edward Wright, the navigation lecturer who wrote the introduction to *De Magnete*, confirmed this date. Gilbert's magnetic philosophy had been 'held back not for nine years only, according to Horace's counsel, but for almost another nine'. Wright knew that Gilbert had not merely sat on his work. He had supplied Gilbert with a lot of information about magnetic navigation, some of which had not been available in 1593, let alone 1583. 'In the meantime', Wright continued, Gilbert went on to 'diligently read and digest whatever had been published'. Moreover, 'after long years at last, by means of countless ingenious experiments', Gilbert had dragged magnetic philosophy 'out of the darkness and dense murkiness ... of incompetent and shallow philosophizers'.

So much, then, for Gilbert the inductive 'Baconian' philosopher of magnetic facts. It seems that the hypothesis came first and the proof later. Gilbert spent the best part of two decades before he published gathering data, designing experiments to test his theory and reading up on magnetism. He studied della Porta's *Natural Magic*, and expanded della Porta's and Peregrinus' investigations into a huge, sophisticated corpus of experiments on spherical and other magnets. He read the slim sections on the lodestone in many philosophical works, and some of the vast corpus of navigational works that also dealt with it. He made the acquaintance of sea captains like Francis Drake, compass experts like William Barlow and practical mathematicians like Wright. As he gathered compass data to support his philosophy, it became clear that the theory of a magnetic Earth promised to revolutionise navigation. By 1597, the utility of magnetic philosophy gave him added reasons to publish. Edward Wright now offered considerable advice, and made sure that the navigational applications were fully explored.

Gilbert judged that *De Magnete* was ready for the press in 1600. He expected trouble from conventional scholars. He was not worried about the fate of his magnetic facts, but of his new magnetic philosophy. Gilbert's book was the carefully crafted defence of magnetic philosophy's central principle – the vital, noble, moving, heavenly, magnetic Earth.

A New History of Science for a New Earth

De Magnete has the status of a classic book because many historians have viewed it as a break with Renaissance natural philosophy and the début of modern experimental science. Gilbert's concept of the Earth's magnetic soul and his *Nova Physiologia* show that it is too simple a view, but there is truth in it. Gilbert had left the Renaissance behind, but his most revolutionary leap has gone unnoticed. It explains why *De Magnete* was the first great experimental book, and why Gilbert was, like Francis Bacon, an advocate of a fresh, experimentally based start in natural philosophy. Gilbert thought that the history of science demanded it.

As the word Renaissance (rebirth) implies, the project uniting Renaissance natural philosophy was the rediscovery, analysis and improvement of ancient learning. Renaissance thinkers did not have our modern concept of progress. For them, progress consisted of escaping from the ignorance of the Middle Age (as they saw it) by emulating the achievements of Antiquity. Some writers before Gilbert and Bacon acknowledged that the compass was a proof that modern 'arts' or technology had achieved more than the Greeks and Romans. But the arts were not philosophy. The basic concepts and systems of Socrates, Plato, Aristotle, Galen, Ptolemy and others still represented the pinnacle of human knowledge of nature, even if there were errors to be corrected and new discoveries to be incorporated.

Influential Christian humanists said that the greater wisdom of the ancients stemmed from the fact that their era was closer in time to God's creation of the world. Before Adam sinned, God had given him a perfect knowledge of nature. It had degenerated along with the general moral and physical degeneration of the world. A corollary of this sacred history of decline was that more ancient philosophers than Aristotle might have been even wiser. That was why Marsilio Ficino revived Platonism, and then went on (or back) to study Pre-Socratics like Pythagoras (who may have existed) and the very ancient Egyptian priest Hermes Trismegistus (who had not). Neoplatonism and 'Hermetic' magic were the boom industries of sixteenth-century learning.

This 'back to the future' ideology did not stop Renaissance mathematicians and natural philosophers making real innovations. Copernicus' heliocentric astronomy is a perfect example. But Copernicus represented his work as the recovery and improvement of a system known to pre-Ptolemaic thinkers, including Hermes. So did Bruno. Patrizzi's theory of matter in his *New Philosophy* was equally new, but he also justified it with references to pre-Aristotelian ideas, which Gilbert called 'tatty ancient fragments'.

Clearly, Gilbert's history of science was different. He called the first chapter of *De Magnete* 'Writings of ancient and modern authors concerning the loadstone: various opinions and delusions'. His opening sentence was a stark challenge to Renaissance ideology.

In former times, when philosophy, still rude and uncultivated, was involved in the murkiness of errors and ignorances, a few of the properties of things were, it is true, known and understood: in the world of plants and herbs all was confusion, mining was undeveloped, and mineralogy neglected.

The lodestone was discovered by chance. Little was known, but Plato and Aristotle still philosophised about it. 'Lest the story of the loadstone be too brief ... [the Greeks and Romans] appended certain falsehoods which in the early time no less than today were by precious sciolists and copyists dealt out to mankind to be swallowed.'

There is more in this vein scattered throughout *De Magnete*, and much more in *De Mundo*. Gilbert had sketched out a complete history of science. The orthodox Renaissance view was that scholars were building upon various ancient traditions of truth. Gilbert's revolutionary purpose was to show that they were all false, and had become more false as the centuries went by.

Gilbert applauded humankind's first efforts in natural philosophy. He praised the ancient Egyptians and pre-Socratics, especially Thales, who suggested that lodestones had souls. But he was no believer in their ancient or revealed wisdom. This was an age of uncivilised ignorance. They did well because their knowledge arose from their practical experience of making nature serve human needs. And they 'philosophised freely, having not pledged themselves in the words of a master, as happens today'.

Disaster arrived with Aristotle, the so-called 'master of those who know'. Aristotle's natural philosophy did not become the dominant paradigm because it was self-evidently true, or even very good. 'Aristotle obtained power and influence over other philosophers just as his pupil Alexander [the Great, who was also his patron] obtained them over the world's rulers.' The triumph of his theory of matter was secured by doctors like Galen, theologians like Aquinas and university professors like the scholars he had left behind in Cambridge.

Finally, once Galen had pursued [Aristotle's matter theory] *and the doctors had added their weight, this entire lower world was seen by an amazing transformation to be arranged into four elements, so very different to the earlier age. Next, everyone had to swear by these words, and then the model was absorbed into the standard way of teaching and discussing religion. It is compulsory for young students to be presented with these trivialities.*

Gilbert also had a parallel history of mathematics and astronomy. It explained why so many accepted that the heavens were made of solid spheres packed around the Earth. Astronomy began in the Middle East as a practical art. People needed predictions of the future courses of the Sun, Moon and planets. The earliest astronomers invented suitable methods of calculation. They conveniently hypothesised that the planets moved in circles. They did not care whether the planets really did so, as long as they got accurate results. Then Aristotle the tyrant declared that the stars and planets did move in circles, because they were borne round by solid spheres. Finally, theologians had made the spheres a virtual article of faith when they made them the natural shapes in God's perfect heavens, and contrasted their circular perfection with the imperfect motions of the corrupt terrestrial world. In short, 'the Mathematicians deceive the gullible Philosophers, and the Philosophers dull the Theologians' senses'.

These novel histories were the work of a true revolutionary, who believed that all the Western traditions of natural philosophy, astronomy and medicine were false. Gilbert wanted to jettison the central principle of Renaissance learning in favour of a modern belief in progress. The ancients were not models for imitation. The moderns were their superiors because 'from then until now men have gradually gained in wisdom' – or would have done if political, sociological and religious factors had not conspired to keep natural philosophy stuck in the dark ages of ancient Greece.

There was one contemporary of Gilbert's with an equally radical post-Renaissance history, Francis Bacon.

After Gilbert's death, Bacon wrote of his belief in progress, and in the tyranny of Aristotle and the schools. He too developed sociopolitical explanations. It would be unfair to repeat that Bacon was one of the few readers of both *De Magnete* and *De Mundo*. Their historical views were but one reason for their promotion of experiment.

· Chapter 9 ·

Gilbert the Experimenter

De Magnete was full of new experiments and discoveries and relied upon them to an unprecedented extent. Gilbert made no pretence of finding classical antecedents to magnetic philosophy. He made no attempts to reconcile it with the concepts and language of traditional theories. There was no point in emulating Renaissance scholars, who looked for truth in the wrong direction.

> [W]e do not at all quote the ancients and the Greeks as our supporters, for neither can paltry Greek argumentation demonstrate the truth more subtilly nor Greek terms more effectively, nor can both elucidate it better. Our doctrine of the loadstone is contradictory of most of the principles and axioms of the Greeks.

To be sure, there is no description of an experimental method to rival Bacon. Gilbert said only that, like geometry, his experiments ascended in a series from well-established facts to 'things most secret and privy in the earth'. He wanted people to repeat them, so he put ...

> ... larger and smaller asterisks against them according to their importance and subtlety. Let whosoever would make the same experiments handle the bodies carefully ... When an experiment fails, let him not in his

ignorance condemn our discoveries, for there is nought
in these Books that has not been investigated and again
and again done and repeated under our eyes.

Gilbert urged those unconvinced by his philosophy to 'enjoy' his experiments nevertheless and 'if ye can, employ them for better purposes'. Facts about magnets and compasses come so thick and fast that enjoyment can elude the biggest enthusiast. Even so, Gilbert's book is not really *'de magnete'*, not 'on the lodestone'. As the rest of the title made clear, it was 'A new natural philosophy of the great lodestone the Earth, demonstrated with many reasons and experiments': not a science of magnets, but a magnetic philosophy.

De Magnete (as we must reluctantly continue to call it) was organised to prove wrong Aristotelians and Ptolemaists. The Earth was not made of 'earth', but a kind of pure lodestone. Gilbert set up lodestone as the antithesis of simple, cold, dry, inert Aristotelian earth. Its magnetic poles and axis showed that it had structure and form. Its attraction of other magnets proved that it was active, not passive. Magnetism was an immaterial force, unimpeded by solid barriers. Immateriality, vitality, structure, self-motion – these were the properties of soul, not matter. Lodestones, then, were signs of the Earth's soul. Finally, small magnets rotated, evidence that the giant terrestrial magnet rotated too.

Theory in Experiments: Gilbert's Terrella and Versorium

Almost all of *De Magnete*'s experiments and arguments

were marshalled around Gilbert's theoretical agenda. His theory not only guided the experiments in *De Magnete*, but also even controlled the design of his apparatus. Philosophers of science discuss 'the theory-ladenness of observation' – that is, the extent to which scientists' theories pre-structure what they observe in their experiments. In Gilbert's case, the extent was undisguised.

The key piece of experimental apparatus in *De Magnete* was effectively Peregrinus' spherical lodestone. This unusual kind of magnet made its entrance in the third chapter. All lodestones have poles but 'the spherical form, which, too, is the most perfect, best agrees with the earth', he stated baldly.

Gilbert knew perfectly well that spherical lodestones were difficult to use, let alone to acquire, compared with bar or oval magnets. These were the shapes familiar to della Porta's readers, and *De Magnete* used them too. In comparison, the all-important magnetic poles and axis of a 'little earth', or terrella, were not obvious. Moreover, it was weaker than a bar magnet.

So which magnet has the 'best' shape for investigating magnetism, especially the Earth's magnetism? The modern choice of bar magnets emphasises modern theory that a magnet consists of atomic dipoles aligned end to end to form a straight axis, where the lines of force are densest. Strong, labelled poles draw attention to the superficially rectilinear motion of iron attracted swiftly towards them. The middle region is insignificant.

But where the lines of force run externally from south to north poles, they are curved and sphere-like. Moreover, weak magnets like the Earth can align other magnets in their field without actually moving them

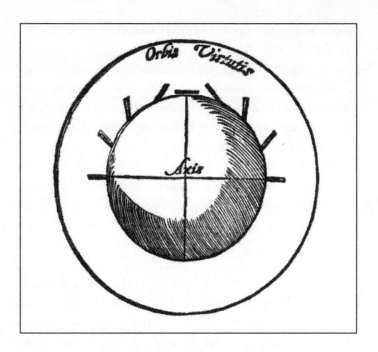

Illustration 6: Gilbert's terrella.
This is a typical illustration from *De Magnete* of the use of a terrella. Gilbert used spherical lodestones because they imitated the Earth. Note the iron wires attached to it, which exhibit on the terrella Robert Norman's recent discovery of the inclination of compass needles. The diagram also illustrates Gilbert's concept of the magnetic *orbis virtutis* [sphere of virtue], extending into space around the terrella. Modern physicists would call this the magnetic field, and say that the iron wires indicate the direction of the lines of force in the field, but Gilbert's was not a primitive field concept.

through space. The force is a torque, rotating the magnet until its axis aligns with the lines of force without moving it along them. Gilbert's spherical magnets emphasised these properties. And when his magnetic spheres moved in a magnetic field, there was no mistaking how they moved: they gyrated like a Copernican Earth.

'To this round stone', he wrote, 'we give the name Μικρογη or *Terrella*.' These are Greek and Latin respectively for 'mini-Earth'. To explore Gilbert's theory-laden terrella experiments was to have followed him half-way to his conclusion.

Almost as important as Gilbert's terrella was his versorium, or 'rotation detector'. It was a pivoted iron needle, essentially a miniature compass. If Gilbert adapted the terrella from natural philosophy, the versorium came straight out of navigation. From Peregrinus to della Porta, natural philosophers had laid little iron wires or filings directly on the surface of their lodestones. Gilbert's versoria rotated freely in three dimensions.

In terms of modern physics, the versorium is a well-designed instrument for detecting the three-dimensional field of a magnet and the torque it exerts. Gilbert did not develop a theory of the magnetic field – his concept of magnetism depended too much on pre-modern ideas of a magnetic soul and magnetic form for that – but he offered something similar. Every magnet was surrounded by an invisible, immaterial 'sphere of virtue'. The movements of the versorium provided visible data with which to map this elusive entity.

Gilbert used his versorium to prove his theory that magnetic matter did not move in straight lines, like elemental earth, but in the circles normally reserved for heavenly bodies. Like the terrella, it was theory laden. When a versorium was brought near a terrella, the little needle's mounting resisted the force of attraction and prevented it from flying straight to the lodestone, emphasising the rotational components. The versorium was well named, although it was not so much a detector as a generator of circular magnetic motions.

Gilbert's versorium did not just provide the magnetic rotations that his theory needed. Its other achievement was to demonstrate that the terrella really was a model Earth. The most convincing of Gilbert's 'principal demonstrations' consisted of moving versoria through the magnetic field of a terrella and recording their precise orientation. For Gilbert this was equivalent to sailors taking measurements of variation or inclination on their voyages.

Much of *De Magnete* was devoted to demonstrating the exact fit of laboratory data to terrestrial observations. He reasoned that the Earth must be a magnetic sphere like the terrella, but also that the terrella was a reliable laboratory model of the Earth. Without leaving his laboratory or risking shipwreck, Gilbert could circumnavigate the globe at will. His surrogate ship and compass, the versorium, had the advantage of going anywhere, fast – even to the polar regions. Gilbert built a systematic observational database that was beyond the combined expertise of the English, Dutch and Spanish navies.

The Central Principle of Analogy

Gilbert's invention of the versorium and re-invention of the terrella provided him with a commanding set of demonstration experiments, but they were only convincing if you accepted a central principle of magnetic philosophy – a central principle of analogy. You had to agree that a spherical lodestone, turned on a lathe or moulded in the laboratory, was indeed a 'little Earth'. A strong corollary of the general analogical principle was that the properties of the part (the terrella), especially its circular motions, were also properties of the whole, the Earth.

For the corollary, Gilbert was on safe ground – Aristotle's ground. Gilbert adopted without question the Aristotelian principle that 'the natural movements of the whole and the part are alike'. We no longer have a concept of 'natural motion', but it was fundamental to Gilbert. Aristotle had argued that the 'natural motion' of Earth was a straight line towards a resting place at the centre of the universe. Gilbert agreed that 'earth' had a natural motion, but it was circular and rotational, just like the natural motion of the planets.

The analogy itself was less Aristotelian. Philosophers drew a fundamental distinction between 'nature' and 'art'. Natural philosophy dealt with the normal course of nature, and reasoned from empirical observations of nature's processes. 'Art' or technology concerned human intervention in nature, and forcing nature to serve human ends. The knowledge of 'artists' could never be natural philosophy, because it did not produce knowledge of the natural course of things. A good example from practical mathematics was 'the art of gunnery', or ballistics. Sixteenth-century engineers developed a sophisticated knowledge of the parabolic trajectory of missiles, but their knowledge was of violent or unnatural motion. Chemists seemed to deal with substances and violent reactions that did not occur naturally.

Gilbert's experiments did not use natural 'parts' that

could just be picked out of the ground. He used smelted, beaten iron needles restrained on a pivot, and stones 'turned on a lathe ... The stone thus prepared is a true homogeneous offspring of the earth'. Were these natural or the products of art? The answer to that question determined whether you accepted the validity of magnetic or many other experiments in the laboratory (where people labour to control nature). For centuries the answer had been 'No'. Late Renaissance philosophers, including Aristotelians, were just beginning to reassess whether 'artificial experiences' such as Gilbert's had any value to natural philosophy. *De Magnete*'s persuasive establishment of the terra–terrella analogy was a turning point in the history of experiment. Today we barely stop to think that the jars of pure chemical elements, the vacuum in an air pump or the pure-bred strains of *Drosophila* flies that we encountered in school science lessons do not exist in 'nature' and could not have advanced traditional natural philosophy.

Gilbert organised *De Magnete*'s experiments and discussions into six books. There was a general survey, followed by one book on each of the five magnetic rotations that he identified.

CHAPTER 10

ANY OLD IRON?

The first book of *De Magnete* opened with an allusion to Gilbert's novel history of scientific error. He launched into his first demolition job on 'all these philosophers, our predecessors [who], discoursing of attraction on the basis of a few vague and indecisive experiments ... are world wide astray from the truth and blindly wandering'.

Gilbert gave grudging approval to a few of the more experimental authors. 'Petrus Peregrinus [wrote] a pretty erudite book, considering the time.' Apart from criticising Peregrinus' polar theory of direction, and questioning his perpetually rotating lodestone, this is Gilbert's only reference to the inventor of spherical magnets. Della Porta was 'worthy of praise', though much of the information, from others 'or through his own studies is not very accurately noted or observed'.

Gilbert actually drew a number of experiments from della Porta's *Natural Magic*, which illustrates the importance of magic to the history of experiment. Some were 'trite and familiar': lodestones have poles, opposite poles attract, a magnet divided becomes two magnets, lodestone attracts iron more strongly than another lodestone. Others were more novel and useful to him.

Gilbert was not impressed with della Porta's causal explanations. *Natural Magic* simply did not discuss why magnets had the property of direction, though della Porta's general theories of correspondence, sympathies, antipathies and celestial virtues had been used by others. Della Porta did have a theory of the magnetic attractions between lodestones and between lodestone and iron. It depended upon the traditional classification of stones and metals as different kinds of stuff. 'I think the Loadstone is a mixture of stone and iron, as an iron stone, or a stone of iron ... Whilst one labours to get the victory of the other, the attraction is made by the combat between them.'

Aristotelian textbooks similarly held that ores consisted of iron, which was primarily 'watery' (which was why metals melted), mixed with stone, which was earthy and infusible. The iron could be separated by the metallurgist's alchemical processes. The occult attraction of lodestone for iron, like other occult sympathies and antipathies, showed, not a substantive identity, but a surprising, hidden relationship between different species, such as the olive and the grape, or lodestone and antipathetic garlic!

Bringing Iron and Lodestones Together

The fact that lodestones were rare, and supposedly of a different species to iron, threatened Gilbert's theory of the magnetic Earth. If lodestones were occult oddities, how could the whole Earth be made of a lodestone-like substance? Gilbert's response was to propose a new theory of metals and stones in which lodestone and iron were basically the same substance. He agreed that lodestone was rare but its brother iron was not.

Their identity followed from the element theory that he developed more fully in *Physiologia Nova*. There was true earth and its related humour. When humour entered the solid matter of true earth, it formed iron and lodestone, with lodestone as 'a noble kind of iron ore'. If it entered the more degenerate earthy solids found on the Earth's surface, it formed other, non-magnetic metals and rocks. Many of his new experiments and discoveries were designed to show that iron was virtually identical with lodestone.

The abundance of iron had just become geologically plausible. Europe's demand for iron grew fast throughout the sixteenth century. The landscape of Elizabethan England had visibly changed in order to meet it. Deep mining technology now sunk shafts far into the Earth to exploit plentiful seams of iron ore, and forests were shrinking fast to supply charcoal for smelting. Iron was getting cheaper, though not yet cheap enough to be discarded and recycled by scrap merchants collecting 'any old iron'. As Gilbert pointed out, iron miners were also unearthing unprecedented quantities of lodestone, but no traces of pure Aristotelian earth!

Gilbert likened the miner's new explorations under the Earth's surface with the navigators' voyages over it. They both confirmed his view that it was the technological progress of practical men, not the growing number of natural philosophy books that was improving knowledge of the Earth. He acknowledged the ambivalent results of technological progress. He listed fifty-eight uses, from swords to ploughshares, plus:

those pests of humanity, bombs, muskets, cannon-
balls and no end of implements unknown to the Latins.
I have recounted so many uses in order that the reader
may know in how many ways this metal is employed; it
is smelted daily; and there are in every village iron
forges. For iron is foremost among metals and supplies
many human needs, and they the most pressing: it is
also far more abundant in the Earth than other metals,
and it is predominant. Therefore it is a vain
imagination of [al-]*chemists to deem that nature's*
purpose is to change all metals into gold.

Iron, of course, was thought of as a 'base' metal. In Aristotelian terms, it was at the first stage of the growth of a metal from potentiality to actuality. The 'noble' metals were the most fully actualised. That was why one could find pure gold and silver, but only in minute quantities.

Gilbert needed to displace this value-laden philosophy of metals with one that made iron more noble. Some Renaissance humanists had already done so. Thomas More made iron more valuable than gold in his *Utopia*, precisely because of its greater utility, and an anonymous English treatise on coinage of 1581 had made the same point. This new stage in the social history of iron was one of many midwives at the birth of our magnetic Earth.

The Magnetic Earth: Theoretical Dream or Empirical Reality?

Gilbert introduced his brainchild at the end of Book I. He had already introduced his terrella, versorium and the analogical argument. Now he laid his cards on the table: 'We must formulate our new and till now unheard-of view of the earth, and submit it to the judgement of scholars.' The Earth is made of the purest lodestone. It is not 'Aristotle's "simple element" and that most vain phantasm of the Peripatetics – formless, inert, cold, dry, simple matter, the substratum of all things, having no activity, – [these] never appeared to anyone even in dreams'.

But nor had Gilbert's superlodestone. We don't find it, Gilbert explained, because the surface of the Earth is degenerate. At the surface, original pure earth had been subjected to the Sun's light and heat for a long time. It was dried out, and profoundly altered. That was why most rocks were not magnetic or even ferrous.

Gilbert's magnetic Earth was a theoretical entity – never directly observed but inferred from the phenomena. There is nothing wrong with that kind of scientific reasoning. Modern science is full of theoretical entities – the magnetic monopole, or even the electron, are excellent examples. But his use of it seriously undermined some of his critiques of Aristotelian matter theory. Aristotelians held that the four elements always occurred in mixtures. Gilbert neglected to mention that.

His protective hypothesis was not unlike our division of the Earth into a large core and thin crust. He conceded that the Earth's surface was made up of heterogeneous and not very magnetic substances.

But the great bulk of the globe beneath the surface and its inmost parts do not consist of such matters; for these things had not been were it not that the surface was in contact with and exposed to the atmosphere, the waters, the radiations and influences of the heavenly bodies.

The claim of the giant lodestone had been made. Now he had to prove it. The remaining five books of *De Magnete* set out five different ways in which the terrella was analogous to the Earth, with a digression on electricity. Each book dealt with one of five circular magnetic movements. The emphasis on rotation set the scene for the climactic sixth book on magnetic revolutions, where he argued that the Earth was rotated daily by its magnetic power. But first came coition, Gilbert's sexy word for attraction; then direction, or the property of verticity that aligned magnets north–south; followed by the motion of variation away from true north; and, in the fifth book, the new discovery of inclination.

· CHAPTER 11 ·

THE BEAUTIFUL UNION OF MAGNETIC SOULS AND BODIES

Book II introduced Gilbert's concept of coition. Coition is, quite literally, a lovely word. It means a 'coming together', like the union of two lovers' bodies. It has resonances of the occult philosopher's principles of concord and strife, but without the strife. Gilbert did not actually think that two magnets made love, but he was convinced that their motions were governed by principles of mutual harmony.

Coition is not Attraction!

Magnetism moved the whole Earth in harmony with its fellow planets and made sure that lodestones were in peaceful union with the Earth and with each other. Aquinas' scholastic explanation of magnetic 'attractio' had a dominating agent seizing and altering a 'patient' magnet before dragging it away. Della Porta's magical explanation derived attraction from strife and discord in the lodestone. If there was one concept that Gilbert wanted to banish, it was attraction.

'Coition, we say, not attraction, for the term attraction has wrongfully crept into magnetic philosophy through the ignorance of the Ancients; for where attraction exists, there, force seems to be brought in and a tyrannical violence rules.' He wanted to replace the

123

Greek term with his neologism but, as history has shown, linguistic habits proved too powerful, even for Gilbert. 'If we have at any time spoken of magnetic attraction [he had], what we meant was magnetic coition.'

Not surprisingly, the violence implicit in the concept of 'repulsion' offended Gilbert as much as did attraction. Indeed, Gilbert claimed never to have observed repulsion between two like poles. Once again, he adapted Peregrinus' experimental apparatus to produce results in accordance with his beliefs. The apparatus is certainly not a staple of the physics storeroom today. But it's not too difficult to observe two magnets coming together in harmony with no apparent signs of violent attraction or repulsion.

Take two conveniently sized magnets. Bar magnets will do, but terrellae are, of course, best. (You can imitate Gilbert's own imitation Earth by hollowing out a core in a spongy ball and stuffing a bar magnet down this 'axis'.) Place the two magnets on floats in a basin of water. Gilbert placed his terrellae on circular discs of wood that had a hemispherical depression in the centre to receive them, but polystyrene is easier to work with. Align the two magnets in any way you like, although the most spectacular effects occur if you oppose the like ('repelling') poles. Hold them still, a small distance apart, until the water in the basin is reasonably still. Let go, and watch.

Both magnets begin to rotate, in order to increase the distance between their like poles and decrease that between their unlike ('attracting') poles. More slowly, because of greater water resistance, they start moving through the water, according to the resultant force of all the repulsions and attractions between the poles. If like poles were opposed, the magnets briefly drift apart. But, since they are rotating, the resultant force changes rapidly, and soon they move together. The angular momentum of each magnet-plus-float is considerable (especially if, like Gilbert, you use lodestones and wood). Thus, even when two unlike poles come close together, the magnets keep on rotating. And so they spin past and around each other until, after some time, they gently come to rest with two unlike poles in close proximity. As Gilbert put it:

If two loadstones be set over against each other in their floats on the surface of water, they do not come together forthwith, but first they wheel round, or the smaller obeys the larger and takes a sort of circular motion; at length, when they are in their natural position, they come together.

This, then, is coition – the mutual action of two magnets to achieve union according to the laws of magnetic philosophy. Gilbert conceded that unmagnetised iron behaved differently – 'it flies to [the lodestone]'. Because it was not a polar magnet, 'there is no need of these preliminaries' or, one might say, foreplay. The magnetic coition of true earth was poles apart from the Aristotelian paradigm of rectilinear motion. Gilbert was sure that magnetism was not just an extraordinary force, but a unique one. It was unique because it was immaterial, or incorporeal.

Electricity is not Magnetism!

It was to establish the uniqueness of magnetism that Gilbert began Book II with his famous experimental investigation of static electricity – and his less celebrated effluvial explanation of it. His little excursion into 'the amber effect' makes up less than 7 per cent of *De Magnete*, but has worked wonders for his historical reputation. The heroic title of 'father of electricity', coined for him in the nineteenth century as electrical machines became commonplace, is still bandied about. The first great historian of Gilbert, Sylvanus P. Thompson, was a member of the Institute of Electrical Engineers (IEE), and the IEE commissioned the painting that adorns this book – the fanciful scene of Gilbert performing electrical (not magnetic!) experiments for Queen Elizabeth.

Gilbert discussed electricity in *De Magnete* only to show that it was nothing like the noble quality of magnetism. He needed to rescue the lodestone from 'the asylum of ignorance' of occult qualities, in which amber was the other famous resident. Gilbert was predictably disgusted: 'In very many affairs persons who plead for a cause the merits of which they cannot set forth, bring in as masked advocates the loadstone and amber.' Galen was one culprit.

Gilbert's strategy was to show that both magnetic and electric 'attractions' were too common and universal to fit the accepted idea of an occult quality. But where magnetism was the property of true earthy substance, many substances were electrics, or what we call insulators. Where magnetism was an immaterial manifestation of the Earth's soul, electricity had a material cause. And where magnetism was an enduring action at a distance, the electric effect was ephemeral and affected by many factors and conditions.

By Gilbert's time, jet had been shown to have the same properties as amber (or 'elektron' in Greek). Gilbert added many more – for example, gemstones, glass, many resins and sulphur. He showed that, unlike magnets, they attracted many substances – not just the classical chaff, but water, many particles, and even metals. That allowed Gilbert to recycle his iron versorium into a kind of electroscope. The phenomenon was so widespread that he invented the name '*electricitas*', which we use today. He certainly baptised, even if he did not father, our modern subject of electricity.

As anyone knows who has tried sticking a balloon to a wall, 'electrics' need to be rubbed before they attract. A common Renaissance explanation of the need for friction was that amber's attraction depended on the quality of heat. Naturally, such a qualitative, Aristotelian explanation appalled Gilbert. He dismissed it with experiments showing that even mild heat destroyed the effect, while magnetic coition survived all but the most intense.

Finally, Gilbert showed how vulnerable electrics were to the material medium through which electricity acted. Damp weather, warm air, not to mention the finest cloth interposed between an electric and a detecting versorium or pile of chaff eliminated it completely. This was a startling contrast with the lodestone, whose powers were unaltered by any medium, even solid rock, except for another ferrous object.

For Gilbert, these results pointed inexorably to a common, material cause for electricity. Electrics were bodies capable of emitting a fine, invisible, but still substantial effluvium. Over short distances the effluvium could surround and move a very light, or pivoted, object until it 'united' with the electric. But the flow was weak, and easily deflected or diffused by many media. We saw that Gilbert thought the cause of electricity was one of his ubiquitous aqueous humours, which some substances released as an effluvium. That was why he concluded his electrical digression with experiments on the surface tension of water. He introduced his effluvial theory of gravitation: 'The earth, by means of the air, brings back bodies to itself; else bodies would not so eagerly seek the earth from heights.'

Readers must have been puzzled by this account, which he only developed in *Nova Physiologia*. But he stated his principal conclusion clearly: 'Electrical movements come from matter, but magnetic from the prime form.' With electricity out of the way, Gilbert could press on to magnetism.

The Soul of the Earth

Matter and form! These were the fundamental concepts of Aristotelian (and Platonic) philosophy that Descartes, Newton and many of the 'new philosophers' of the seventeenth century discarded. Gilbert's works contain strong traces of many other Aristotelian concepts (sphere of virtue, elemental earth, natural motion, humours and so on) that also went into decline shortly after his death.

Concepts such as matter, form, motion and soul are part of metaphysics. Metaphysics provides the abstract philosophical frameworks that pre-structure natural philosophy and all forms of science, including modern physics. Gilbert was revolutionary in many ways, but not in his metaphysics, which were left stranded by Descartes. He protested, adapted and innovated, but his metaphysics belonged to the Renaissance, and therefore owed more than he recognised to Aristotle. Gilbert would have been stunned to learn that, only twenty-nine years after he published, a clever Jesuit produced a brilliant, Aristotelian interpretation of *De Magnete*.

The least modern aspect of *De Magnete* was Gilbert's conceptualisation of magnetism as a form. The properties of lodestones, iron and the Earth itself were produced by their magnetic form. What kind of form? Where did it stand in a hierarchy of forms? Was it material? Clearly not, although electricity was. Was it where Aquinas put it, between material and animate forms? Or was the magnetic form truly a soul? To give the Earth a soul was even more radical than to give it an orbit around the Sun.

In several places Gilbert was as hesitant, evasive or unclear about the Earth's magnetic form as he was about its annual orbit. It was 'as it were, animate' or 'soul like'. But elsewhere he was decisive. His desire to restore to the Earth the dignity that Aristotelians denied it pushed him all the way. He did not jump into the Neoplatonic camp, and make the Earth part of their universal world soul. His Earth had a soul all of its own. In Book V, chapter XII of *De Magnete*, Gilbert let 'Aristotle and his followers' feel his full rhetorical force.

Aristotle concedes to the spheres and heavenly orbs (which he imagines) a soul, for the reason that they are capable of circular motion and action and that they move in fixed, definite tracks. And I wonder much why the globe of the earth with its effluences should be by him and his followers condemned and driven into exile and cast out of the fair order of the glorious universe, as being brute and soulless. ... Thus Aristotle's world would seem to be a monstrous creation, in which all things are perfect, vigorous, animate, while the earth alone, luckless small fraction, is imperfect, dead, inanimate, and subject to decay. On the other hand, Hermes, Zoroaster, Orpheus, recognise a universal soul. As for us, we deem the whole world animate, and all globes, all stars, and this glorious earth too, we hold from the beginning by their destinate souls governed and from them to have the impulse of self-preservation.

It is no exaggeration to say that the main task of *De Magnete* was to prove his startling thesis. What it lacked in metaphysical analysis of the nature of form and soul, and in methodological sophistication, it made up for with wave after wave of new discoveries, ingenious experiments and brilliant syntheses of laboratory and navigational results. Like all Gilbert's best proofs, they depended upon the terra–terrella analogy. Analogy can be an unreliable form of reasoning, especially if the analogy is as radical as Gilbert's. Galileo was not the only seventeenth-century philosopher to criticise his lack of rigour. Gilbert had rejected 'paltry Greek argumentation' for experimental demonstrations. He expressed a rather crude, 'no-nonsense' faith in empiricism when he wrote:

If among bodies one sees aught that moves and breathes

*and has senses and is governed and impelled by reason,
will he not, knowing and seeing this, say that here is a
man or something more like a man than a stone or stalk?*

The lodestone acted like the Earth and, while neither
actually breathed, Gilbert believed that they exhibited
many properties of animated bodies.

Gilbert singled out several signs of the presence of an
animate form: self-motion, immaterial action, internal
organisation, rotation (like that of other planets) and the
apparently purposeful generation of cosmic harmonies.
We saw earlier that these were accepted properties of
souls in the Renaissance. The rest of Book II of *De Magnete*
began to marshal his evidence that magnets had similar
properties.

There is little point in condemning Gilbert for his
belief in the Earth's magnetic soul, and wishing that he
had stuck to magnetic experiments. If he had not become
convinced that geomagnetism proved the Earth's nobility and Aristotle's error, he would never have researched
and written *De Magnete*. Let's just enjoy the ride as
Gilbert brings us into coition with the Earth's 'astral
magnetic mind'.

Experimental Proofs of the Magnetic Soul

The first property of souls, self-motion, was precisely
what Gilbert meant by coition. Whatever the principle
was that moved magnets so mysteriously, and sometimes so beautifully together, it was not an external
force. It resided in the magnet itself. But attraction alone
did not prove the presence of a soul. After all, the

principle of electric attraction was material. The first task was to confirm that magnet coition was immaterial.

Gilbert quickly reviewed and dismissed previous theories of attraction: the atomists' emission of magnetic particles, other effluvial explanations, Cardano's heat, celestial sympathy, Aquinas' substantial form. 'As for the causes of magnetic movements, referred to in the schools of philosophers to the four elements and prime qualities, these we leave for cockroaches and moths to prey upon.'

Previous philosophers had collected evidence that supported an immaterial cause. Della Porta had already demonstrated in *Natural Magic* that magnetism 'can be hindred by no hindrance' of wood, stone or metal except iron. Lodestones magnetised iron through the same media. Gilbert expanded della Porta's trials. He also paid attention to the medium of air, which allowed him to draw out the contrasts with electric attraction. Altering the air between two magnets by heating it, or even interposing a flame, made no difference to coition. Nor did lodestones have to be made to emit effluvia before they attracted. The finest balance detected no weight change in iron after magnetisation. There was no detectable movement of matter out of a lodestone, or in any medium between it and iron.

He showed that really hot magnets (above the Curie temperature, we would say) lost their magnetism: iron temporarily, lodestone permanently.

But fire destroys in the loadstone its magnetic qualities, not because it plucks out of it any particular attractional particles, but because the quick, penetrating flame deforms it by breaking the matter up; just as in

the human body the soul's primary powers are not burnt,
though yet the burnt body remains without faculties.

More evidence of magnetism's immateriality was its well-known capacity to act beyond the boundary of its physical body. We call this the magnetic field. Gilbert called it the 'sphere of virtue'. Gilbert turned the sphere of virtue into an experimental subject with his versoria. He showed how its shape depended on that of the magnet itself, although he concentrated on truly spherical ones. He observed how the 'sphere of influence' extended further than the 'sphere of movement', because versoria aligned at a distance where even the smallest piece of iron was not attracted. He explored how its strength varied with the strength of the magnet itself and distance from the magnet. The force–distance relationship for magnets is complex. It defeated Gilbert and many subsequent investigators. He concluded that it was directly proportional, just as strength, for identical lodestones, was directly proportional to mass.

After immateriality, our third property of a soul was its capacity to create structure and organisation in the matter it informed. All inanimate things except lodestone seemed to be homogeneous in their properties. Gold was equally lustrous in every part. Glass transmitted light in all directions. Any fragment of a piece of amber could attract chaff to any point. But lodestones had poles, near which iron was attracted more strongly and versoria turned faster, in bigger arcs. Lodestones were not unlike plants. Plant grafts had to be inserted in the right direction to respect the internal structure of both graft and host. That was a sure sign of the organising plant soul. But the polarity of lodestones was similar. To make a chain of bar magnets, or to reassemble a broken lodestone, the poles had to have the correct orientation.

All magnets exhibited the same structure. All had their own poles, axis, centre and equator. So did the Earth but, as we saw, they were traditionally conceived as fictions projected by mathematicians on to the Earth from the heavens. It was only the heavens that possessed real, physical poles and an axis, because only the heavens physically rotated. Gilbert the Copernican wanted them to be real terrestrial properties. As a natural philosopher, he took delight in showing that they did indeed have a physical cause, the Earth's animate magnetic form.

Gilbert was anxious to show that the polar properties did not dwell in the parts themselves, perhaps because the magnetic substance was different there. The parts of a magnet depended upon the whole mass, and the form that organised the whole.

Gilbert's analysis of magnetic poles can therefore seem odd to the modern reader, who thinks of them as the ends of a dipole. He imagined that the magnetic power of 'all the parts [of a lodestone], being united in the whole, direct their forces to the poles', under the direction of the magnetic form. This was why a terrella attracted weakly at its equator, not because there was no power there ('in all [parts] the power is in some sense equal'), nor because, as we would say, the separate attractions of the north and south poles cancelled each other out.

In some ways the organising capacity of the magnetic form was even more extraordinary. When bar magnets were strung together, the coition resulted in one body, one long, strong bar. The form even turned iron dust literally into a coherent body. When a piece of a terrella was broken off, it immediately exhibited all the magnetic attributes, including strong poles, even if it had come from the weak equator. The fragment's form had immediately reorganised the magnetic virtue within it.

In a later experimental series that he adapted from della Porta, Gilbert cut through some terrellae. From one he removed everything above its 'Arctic circle'. There was still a north pole, now on its flat top. The equator had shifted slightly south, by an amount exactly proportional to the mass removed. Likewise, removing part of its eastern hemisphere shifted the poles and axis westward by a predictable distance. Half of a terrella divided from pole to pole along its old axis acquired a new axis and poles. No matter what one did, the form conserved the structure of the whole.

Book II contained most of Gilbert's experimental discoveries. Over half of his 180 asterisks were in it. He brought into coition lodestones of many strengths, iron of many shapes. He put iron needles into coition with south poles, north poles, singly and together, in a *karma sutra* for magnetic philosophers. He showed how pieces of unmagnetised iron, when in coition with a lodestone, became magnetic and extended the sphere of influence around it. Like compass makers, he 'armed' the poles of lodestones with iron caps – magnetic viagra to increase their potency and to make needles stand more erectly upon them. The experiments confirmed that magnetic attraction was coition – the mutual self-motion of two bodies informed by the same extraordinary, immaterial, organising, soul-like principle.

Chapter 12

Magnetic Rotations and Navigation: Direction, Variation and Inclination

Gilbert's motion of coition is the most striking magnetic property. It was the main property for previous natural philosophers like Aquinas and della Porta. But it was the least important for proving that the Earth was magnetic. Coition could be observed between lodestones or iron rods, but not between a magnet and the Earth. Compass needles or iron filings do not migrate from the lab bench to Greenland or Antarctica! We would say that the Earth's magnetic field is too weak. If you suitably suspend a magnetic dipole of low mass, such as a compass needle, it will experience a magnetic torque strong enough to align with the Earth's lines of force, but not a force of attraction strong enough to move along the lines of force towards a pole. The weak rotational force was what interested Gilbert. In the next three books of *De Magnete*, he mapped and explained it, and created a whole new 'field' of magnetic science.

The alignment of a magnet with the local lines of force is really one motion. In the case of the Earth's field, we conveniently divide it into a horizontal component (variation) and a vertical one (inclination), owing to the deeply buried position of the dipole. Following Robert Norman's discovery of inclination, Gilbert had a similar model.

He lifted his demonstration right out of Norman's *Newe Attractive* of 1581. Norman had suspended an iron wire in water. The alignment of the one wire simultaneously exhibited variation and inclination, but it was not attracted. Gilbert even dared to add an asterisk to his 'demonstration of the absolute conforming of a magnetic body to unity with the Earth's body'. It was the whole body and not just the poles. Unlike modern physicists, Gilbert located the Earth's poles on its surface, where they were on his terrellae. The experiment confirmed his principle that magnets responded to the Earth as a whole, not to the pole, centre or any other single region. The principle fitted with his concept of the wholistic action of animate forms, but it made the analysis of magnetic motion more complex.

Nevertheless, he did analyse it in two components. He addressed the horizontal motion of direction in Book III. By direction, he meant the property of a small magnet to rotate until its axis aligns with the poles of a big magnet. Norman's vertical inclination appeared in Book V. This too was a rotation. (Gilbert declared that variation, the subject of Book IV, was not a proper motion at all.) These rotations established the fourth property of a soul, circular motion.

Direction and Verticity

Magnetic navigation had turned the lodestone's directional property into a more urgent, if not more puzzling, phenomenon than attraction. The common herd's theories of celestial influences and magnetic mountains did not detain Gilbert for long. He set about proving his theory that the power of direction came from the whole Earth. To do so he needed a new theoretical term. 'Verticity' would replace 'direction', as 'coition' had replaced 'attraction'. Gilbert continued to speak of magnetic poles rather than vertices, but verticity (like polarity, in fact) contained theory-laden resonances of turning, around a pole or in a vortex.

He defined verticity as the 'power distributed by the innate energy [of a magnet] from the equator in both directions to the poles'. The definition immediately excluded Eurocentric explanations based on the influence of a single, extraordinary, always northern, and often extraterrestrial, source. Norman's respective point would not do for the same reasons. Moreover, verticity was possessed by any laboratory magnet. No voyage to Siberia, to the interior of the Earth or even to Ursa Minor was needed to investigate its origin. Gilbert discussed experiments with ferromagnets of many kinds – but behind all the experiments, sometimes implicitly, stood the mother of all magnets and the origin of all verticity, the Earth itself.

To demonstrate the literally mundane and tractable nature of verticity, Gilbert sent the good ship *Versorium* off on the first of its voyages across the terrella early in Book III. Circumnavigators (here Gilbert name-dropped Drake and Cavendish) had recently established that in the explored regions of the Southern Hemisphere 'the lily of the mariner's compass ever points north. ... Our terrella teaches the same lesson.' The terrella showed that it was a universal law – almost. At the South Pole itself, inclination was 90°. It therefore attracted the south-seeking crotch so directly that the north-seeking

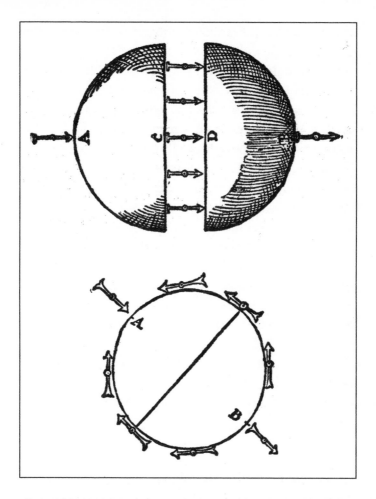

Illustration 7: Versoria indicating magnetic direction in a terrella.

Gilbert's versorium was an adapted compass needle, able to turn horizontally or vertically. In the upper diagram, the versoria are constrained to indicate only direction towards the poles. Note that the tip of the needle always points to the north pole A, except when directly over the south pole B. The divided terrella proved to Gilbert that direction, or verticity, also existed inside the stone.

lily stood upright and away from north. Any sailor who found himself in deepest Antarctica was going to have compass problems to add to his hypothermia and woeful sense of direction.

The terrella also allowed Gilbert to investigate verticity in the interior of a magnet, something quite impossible for the Earth itself. He divided terrellae across various parallels of latitude, including the equator, and put versoria between them. The direction was always from north to south pole. Combining and adapting the two diagrams gives a similar result to maps of the Earth's magnetic field based on the simple dipole model.

From this point in *De Magnete*, Gilbert shifted from a general investigation of magnets to a focus on the Earth. Gilbert's use of the terra–terrella analogy shifted with it. In the first voyage of the *Versorium*, Gilbert drew upon reliable knowledge of the Earth (the north-seeking behaviour of compasses) to secure the credibility of terrella as a laboratory model of the Earth. But now he cashed in that credibility by arguing from the verticity of laboratory magnets to the verticity of the Earth.

Some of his most important investigations were about how unmagnetised (that is, manufactured) iron acquired verticity from a lodestone. Gilbert had a simple test for the presence of verticity in iron. Did the iron rotate into alignment when it was suspended near a stronger magnet? It allowed him to explore very weak magnetisation, the only kind produced by the Earth.

He showed that iron acquired verticity from a lodestone. In iron, the verticity was easily changed. Bring one end of an iron wire up to the south pole [in his terms] of a lodestone. It does not even have to touch. Take it away,

140

and it will point south. Rub the other end and the direction reverses. Repeat as desired. 'Thus you will be able to alter again and again the property of the iron.'

Then Gilbert did away with the lodestone altogether! He was still able to alter the verticity of manufactured iron at will. How?

> *Get a smith to shape a mass weighing two or three ounces, on the anvil, into an iron bar one palm or nine inches long. Let the smith* [you are a gentleman, remember?] *stand facing the north, so that as he hammers the red-hot iron it may have a motion of extension northward.*

Bars produced like this acquired verticity – they passed the test by pointing north! Even hammering was not needed. Just make a bar white hot. That destroyed any pre-existing verticity. Then lay it along a north–south meridian and, when it is cool, it has verticity. Heat it again and turn it though 180° before it cools. You have reversed the verticity! In fact . . .

> *. . . let us see also what position alone, without fire and heat, and what mere giving to the iron a direction towards the earth's poles may do. Iron bars that for a long time – twenty years or more – have lain fixed in the north and south position, as bars are often fixed in buildings and glass windows – such bars, in the lapse of time, acquire verticity.*

We can believe Gilbert when he wrote that 'this seemed to us at first strange and incredible'. In *De Magnete*, it became powerful evidence that the Earth did indeed have the properties of a lodestone. It too had verticity. It could pass that verticity to iron and cause it to rotate into magnetic alignment. And if the Earth had verticity, then it had a magnetic form and magnetic poles.

For Gilbert, of course, the poles were not especially magnetic, but products of the magnetic form. It was not ...

> *that the earth's pole, that identical point* [to the pole of a lodestone] *lying thirty-nine degrees of latitude, so great a number of miles away from this City of London, changes the verticity, but that the entire deeper magnetic mass of the earth which rises between us and the pole, and over which stands the iron – that this, with the energy residing in the field of magnetic force, the matter of the entire earth conspiring, produces verticity in bodies.*

If the poles alone produced verticity, then compasses might reasonably be expected to point straight at them. Gilbert was sure they did not. The 'matter of the entire earth' provided him with his theory of variation.

Variation: From the Ideal to the Real World

Today we think that the major cause of variation is the tilt of the Earth's dipole. Gilbert's reference to the 'earth's pole' reminds us that he held the geographical and magnetic poles to be identical. His explanation was not ours. With very acute hindsight, it is just possible to discern, amid the noise of inaccurate, undated, contra-dictory variation data, a signal coming from a tilted dipole. We saw how some sixteenth-century magnetic longitude schemes were consistent with a tilted dipole model. Magnetic island hypotheses also implied that compasses pointed regularly to somewhere other than the true North Pole. Given the inconclusive data, all of Gilbert's evidence for the terra–terrella analogy was consistent with both a tilted and an axial dipole model.

It is obvious why Gilbert never contemplated a theory of separate magnetic poles. The Earth's magnetic axis had to be the same as its geographical axis if magnetism was to account for the Earth's noble Copernican motions. Cosmological values lay at the core of his magnetic philosophy. He did not intend to let the nautical inconvenience of variation become his cosmic nemesis (although variation would contribute to the decline of magnetic philosophy in the 1640s). He needed to explain it. His theory, expounded and justified in Book IV of *De Magnete*, was an inspired, inspiring, flawed masterpiece. He began:

> *So far we have been treating of direction as if there were no such thing as variation; for we chose to have variation left out and disregarded in the foregoing natural history, just as if in a perfect and absolutely spherical terrestrial globe variation could not exist.*

This was quite an admission, and an omission that must have taxed the patience of any Latinate navigator reading *De Magnete*. All the persuasive experiments with terrellae, all the laws and analogies between the little earth and big Earth, it turned out, applied to an idealised, non-existent world. Gilbert had chosen to imagine away the very same aspect of geomagnetism that had led George Best to declare that 'it passeth the reach of

143

natural philosophy'. His explanation had to be good – and it was. At last, in Books IV and V, navigators would discover why Edward Wright had drawn their attention to a work of wacky natural philosophy.

Gilbert developed his explanation of variation within two tight constraints. One he imposed upon himself: the Earth's magnetic axis was also its axis of rotation. This condition ruled out most existing explanations. His second constraint, the navigators' corpus of variation data, ruled them all out, or did so when Gilbert interpreted them through the sceptical eyes of English experts. He adopted from the experts he consulted the English conviction that variation was real, often large in high latitudes – and fundamentally irregular.

The irregularity disproved all theories that magnetic direction was a single movement towards a mountain, rock, star or celestial pole. If they were correct ...

> ... in different places on land and sea the variation point would in geometrical ratio change to east or to west, and the versorium would always regard the magnetic pole, but experience teaches [that ...] variation changes in different ways erratically ... in different meridians and even in the same meridian.

Gilbert's prime counter-example of irregular changes in variation was typically English: observations made by navigators running the latitude 'from the Scilly Isles bound for Newfoundland'.

De Magnete was a rich vein of variation data, revealing a knowledge far beyond any previous work of natural philosophy. It had chapters on variation in the Atlantic,

Pacific and Mediterranean Oceans, even Novaya Zemlya, the large island in the Arctic Ocean. Since it was a philosophy of the Earth rather than a sailing manual, it also covered 'the interior of the great continents'. The chapters declared to the whole learned, Latinate world what English navigators knew. There was irregularity everywhere, many observations were unreliable, the Iberians' true meridians did not exist, 'variations are greatest in regions nigh to the poles ... and there, too, the changes of variation are sudden, as Dutch observers noted some years ago'.

Gilbert also used the data to attack magnetic longitude schemes. He understood that the idea offered 'a welcome service to mariners and would advance geography very much' but it was 'deluded by a vain hope and by a baseless theory'. There was hope, though. Gilbert had read a translation of Simon Stevin's 1599 work *De Havenvinding*. In chapter XI, he endorsed Stevin's empirical scheme of finding port by matching onboard observations with tabulated but irregular data. It was 'of great importance, if only fit instruments be at hand wherewith the deviation may be positively ascertained at sea'. Chapter XII was a description surpassing that of any navigation manual of how to make and use a variation instrument. Gilbert was none too scrupulous about acknowledging sources, but chapter XII was something else. Gilbert did not write it. However, he had constructed a brilliant theory of variation around the English consensus of irregularity, Stevin's havenfinding scheme and his own theory of the Earth.

Recall that in Gilbert's theory, the Earth was basically a perfect sphere of true magnetic Earth, but its crust had degenerated under the influence of sunlight and other celestial virtues. Aqueous humour had been separated from true earth to form oceans, leaving a residue of non-magnetic rock. Moreover, the Earth's own heat, vulcanism and other geological processes had created continental landmasses and ocean deeps. All these deviations from a homogeneous sphere acted as perturbations upon the Earth's otherwise uniform verticity.

> [T]*he globe of the earth is at its surface broken and uneven, marred by matters of diverse nature, and hath elevated and convex parts that rise to the height of some miles and that are uniform neither in matter nor in constitution. . . . For this reason a magnetic body under the action of the whole earth is attracted toward a great elevated mass of land as toward a stronger body, so far as the perturbed verticity permits or abdicates its right.*

No wonder, then, that variation was irregular. In fact, seas and continents were the superficial signs of greater, subterranean features.

> *The variation takes place not so much because of these elevated but less perfect parts of the earth and these continental lands, as because of the inequalities of the magnetic globe and of the true earth substance which projects farther in continents than beneath the sea-depths.*

A great theory, but could Gilbert prove it? Geological surveys were, of course, beyond him, but he could do *micro-*geology with his terrella. It was time to send the *Versorium* on some more voyages. Once again, Gilbert attributed to the Earth new phenomena discovered in the laboratory that had never been suspected to exist in the Earth.

This thing is clearly demonstrated on the terrella thus: take a spherical loadstone imperfect in any part or decayed (I once had such a stone crumbled away at a part of its surface and so having a depression comparable to the Atlantic Ocean); lay on it bits of iron wire two barleycorns in length, as in the figure. AB is a terrella imperfect in parts and of unequal power on the circumference; the needles E, F do not vary but regard the pole straight, for they are placed in the middle of a sound and strong part of the terrella at a distance from the decayed part: the surface that is dotted and that is marked with cross-lines is weaker. Neither does the needle O vary, because it is in the middle of the decayed part, but turns to the pole just as off the Western Azores. H and L vary, for they incline toward the sound parts.

A similarly impressive experiment used a sound lodestone turned into a sphere but with 'a considerable part of its surface projecting a little above the rest'. Versoria deflected slightly from the poles towards these model continents.

Big geological features produced small variations on the terrella. This was another strike against the theory of a magnetic mountain. However, Gilbert himself needed to explain why variation was greater in high latitudes. His answer exploited the magnetic form's balancing of coition and verticity. Coition was weak at the equator and strong at the poles. Conversely verticity was strong

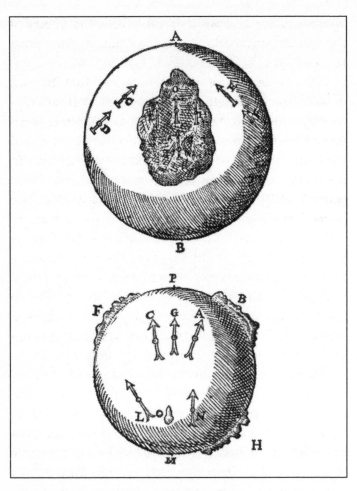

Illustration 8: Magnetic variation on terrellae.
These are classic illustrations of Gilbert's use of his central
principle of analogy between terrellae and the Earth.
The first lodestone was defective, with a 'depression
comparable to the Atlantic Ocean'. The second appears
to have been cleverly turned on a lathe to leave imitation
continents. In both cases Gilbert showed versoria
deviating from the pole in patterns that replicated the
variation of real compasses at sea.

at the equator and weak at the poles. The effects of geological perturbations were greater in the polar regions.

Gilbert's theory of variation shows that he was moving towards a method of analysis and synthesis that created an 'ideal, real world'. Galileo and Newton developed it much more spectacularly when they constructed their laws of motion – laws that do not strictly hold in the messy real world of resistant matter. Gilbert knew perfectly well that neither the Earth nor his terrellae were perfect spherical magnets, but this was the domain in which his analysis of the components of magnetic virtue took place. The real world was generated by synthesising the components of ideal verticity and perturbation. It must be said that Gilbert was no Galileo – as Galileo himself suggested. Gilbert's explanation of the real world of variation was largely qualitative and certainly imprecise.

The nemesis of Gilbert's theory was not methodological, but empirical. It lurked in chapter III, 'Variation is constant at a given place'. Gilbert had no reason to doubt what Stevin, Wright and every expert before them had taken for granted. Indeed, his discovery of the stable magnetic laws underwritten by the Earth's form offered unprecedented assurance that it was true. He had only one caveat. If variation had geological causes, then it was vulnerable to geological catastrophes. But Gilbert was sure it would be 'forevermore unchanging, save that there should be a great break-up of continents and annihilation of countries, as of the region of Atlantis, whereof Plato and ancient writers tell'.

Gilbert could have used a less catastrophic version of

this geological principle to cope with small, local changes in variation. The discovery of secular variation in (and for) London was announced in 1635. By 1650, it had been confirmed in many parts of the world. Gilbert's theory was holed below the waterline. As it sank, new theories based on a tilted dipole were explored. The Siren call of regular magnetic longitude schemes became louder. And once the magnetic poles were free to wander away from the geographical poles, Gilbert's magnetic Copernicanism sunk too.

Nevertheless, Gilbert's philosophy of the magnetic Earth revolutionised magnetic navigation. His accounts of direction and variation became the new consensus, integrating as never before causal explanation and nautical practice. They provided a theoretical justification for the only sensible magnetic longitude scheme, Stevin's havenfinding art. 'Thanks to this magnetic indication', Wright trumpeted (in his 'laudatory address concerning these books on magnetism'), 'that ancient geographical problem, how to discover the longitude, would seem to be on the way to a solution'. And they even influenced voyages of exploration, because Gilbert summarised his irregularity theory into a 'rule of the variation'.

Needles were 'attracted towards a great elevated mass of land'. Gilbert wanted English explorers to concentrate on finding a north-east, not north-west, passage to the Indies. 'Since [in Novaya Zemlya] the compass has so great an arc of variation to the west, it is evident that no continent stretches for any great distance along that whole route eastward.' Of course, the theory, rule and application all proved wrong. Jacobean searchers for the

North West Passage were puzzled to find that it didn't seem to work in Canada either. But counter-instances could be explained away. It was subterranean, not surface, geology that really made the difference. Gilbert's theory was almost irrefutable, but not quite. Time, in the form of secular variation, eventually caught up with it.

Inclination: the Latitude Found!

Variation had been a digression from true magnetic rotations. Book V of *De Magnete*, 'Of the Dip of the Magnetic Needle', brought Gilbert's experimental study of the magnetic form to a climax. Although it was the shortest book, it contained Gilbert's most powerful evidence of the Earth's soul, and the most astounding navigational application of magnetic philosophy.

De Magnete established inclination as a truly global, law-governed and useful property of the magnetic Earth. Gilbert had earlier given credit to 'Robert Norman, skilled navigator and ingenious artificer, who first discovered the dip of the magnetic needle', even if Book V never distinguished between Norman's innovative work and Gilbert's developments of it.

No modern science journal would have accepted Norman's claim of 1581. Granted, compass experts in many parts of the world knew that needles became 'unbalanced' when they were magnetised, but Norman had inferred his general terrestrial cause, the 'point respective', from experiments in only one location, London. Even when *De Magnete* was published nineteen years later, Norman's measurement of 71.5° remained the only published value. Book V therefore began with a description and instructions for the use of a fine inclinometer.

In his analysis of variation, Gilbert's problem had been an excess of data for the Earth, which he had struggled, successfully, to replicate on laboratory terrellae. When it came to inclination, even Edward Wright could not supply him with data from abroad. But Gilbert did not need to wait for navigators to answer Norman's request for more observations. He could supply the world with all the inclinometric data it needed from his laboratory. Book V reported on the surrogate circumnavigations of his versorium.

Inclination was the vertical component of the Earth's magnetic soul. Gilbert had therefore sent his versorium to chart inclination in his laboratory equivalent of the interplanetary space around it! It sounded like an incredible voyage, but he expected his readers would now be convinced of the terra–terrella analogy. Anyone possessing the basic apparatus of experimental magnetic philosophy could decide whether or not it was a flight of fancy.

In chapter II, Gilbert described 'the dip of a terrella representing the earth relative to the standard representation of the globe of the earth, at north latitude 50 degrees.' 50°N was a shrewd choice, because that was London's approximate latitude. The value on the terrella accorded with Norman's.

Values elsewhere quickly refuted Norman's respective point theory. Inclination was 0° at the terrella's equator. As the versorium voyaged to the north pole, it increased in a smooth but non-linear relationship to 90°. The relationship meant that lines drawn by projecting the axis of the versorium converged on no point whatsoever, not the centre of a magnet, its poles nor any 'respective' point. The lack of a simple cause pleased Gilbert. It was more proof of the complex organisation of matter by the magnetic form. Indeed, it proved that the form was a proper physical cause of geographers' parallels of latitude, as well as the poles and axis.

The fact that there was no simple mathematical relationship on a terrella between latitude and inclination did not inhibit Gilbert from his stupendous announcement that magnetism provided an alternative to astronomical methods for finding latitude. The correlation was empirically established with ease on a terrella. With good inclinometers navigators could do the same on the Earth.

In his 'address', Wright's enthusiasm was only a little tempered by empirical caution. He thought it '(to say the least) highly probable' that an observation-based method would work on the Earth. '[T]hus the dip being carefully noted and the latitude observed [astronomically], the same place and the same latitude may thereafter be very readily found by means of a dip instrument even in the darkest night and in the thickest weather.' Gilbert was more rhapsodic. In contrast to other natural philosophies:

[w]*e can see how far from idle is the magnetic philosophy; on the contrary, how delightful, how beneficial, how divine! Seamen tossed by the waves and vexed with incessant storms, while they cannot learn even from the heavenly luminaries aught as to where on earth they are, may, with the greatest ease gain comfort from an insignificant instrument.*

Gilbert was excited because *De Magnete* cracked open the complex mathematics of the dip–latitude relationship. *De Magnete* contained numerous quantitative experiments, but it rarely analysed phenomena geometrically in the manner of Galileo or Newton. The impressive, asterisked geometrical construction discussed in chapter VII is therefore something of a surprise.

The chapter explained how inclination for any point N, of latitude $x°$ on the surface of a spherical magnet

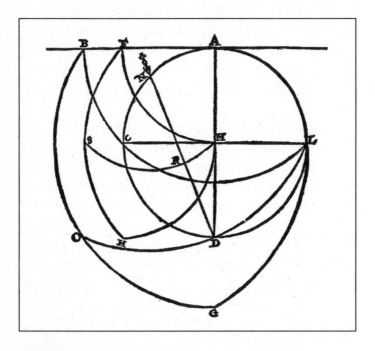

Illustration 9: Geometrical construction of inclination on a terrella.
This is the most fascinating use of mathematics in *De Magnete*, a book that gave contradictory impressions about its role in natural philosophy. See the accompanying text for its significance.

could be determined by constructing a series of arcs. [Readers with an aversion to geometry do not have to follow the construction too closely!] The important arcs were BOG and ODL, and the purpose was to find the point D. The radius of BOG was MB, a quantity determined by the chord AC running from the magnet's pole C to point A on the equator. It therefore lay beyond the physical body of the magnet. The radius of ODL was NL, where L was the opposite pole to C. It began at L and was extended until it cut BOG. ODL was then divided into 90 equal parts (for the 90° of latitude). D was the point where OD was *x* parts long and DL 90-*x*. A line drawn from N to D was supposedly the direction taken by an inclinatory needle placed at N. In this (poorly chosen) example D happens to coincide with the magnet's equator, but only because N has a latitude of 45°.

The construction, technically a kind of nomograph, generates inclination values remarkably well. It was the basis of tables published in the 1600s. Its complexity and apparent arbitrariness suggest that considerable geometrical ingenuity and trial and error went into finding it. Like other nomographs, the diagram seems to be a good trick for getting the right empirical answer, rather than a mathematical model of the causes of inclination. Mathematicians were good at devising tricks that ignored the physical causes of things. As Gilbert had pointed out, it was what Renaissance astronomers did when they invented their 'theorics' to predict the motions of planets.

The arc BOG, which Gilbert called 'the terminus of the arcs of rotation', is especially odd. In three dimensions it formed a sphere surrounding the magnet at a seemingly arbitrary distance from it, and yet it defined the inclination of a needle on its surface. How could it be a real cause? Modern readers of *De Magnete* have understandably thought that the method is a non-causal nomographic device. Yet in his history of mathematics, Gilbert deplored the way that 'mathematicians deceive the gullible philosophers' into thinking that mathematical devices such as the celestial poles or planetary spheres were real.

The Magnetic Soul Demonstrated!

Gilbert did think that his spectral magnetic sphere was real. This was the most important outcome of his work on inclination. When he moved a versorium around a terrella at some distance from it, as though over the surface of a larger but immaterial magnetic sphere, the dip–latitude relation still held. But now the versorium orientated itself towards immaterial points in the terrella's extended sphere of virtue.

In the penultimate chapter, 'Of the formal magnetic act spherically effused', Gilbert too was effusive. Having

labored long and hard to get at the cause of this dip, we have by good fortune discovered a new and admirable science of the spheres themselves – a science surpassing the marvels of all the virtues magnetical. For such is the property of the magnetic spheres that their force is poured out and diffused beyond their superficies spherically, the form being exalted above the bounds of corporeal nature; and the mind that has diligently studied this natural philosophy will discover the definite causes of the movements and revolutions.

... In these spheres (and they may be imagined as infinite) the magnetic needle or versorium regards its own sphere in which it is placed and its diameter, poles and equator, not those of the [physical] *terrella.*

Gilbert had turned Norman's novelty into his best proof of the Earth's soul. What else was capable of existing without matter, as did the effused magnetic sphere? What else was capable of producing and directing a rotation as divine as dip? What else could organise all the parts of a body to co-operate holistically?

Gilbert called the twelfth and final chapter of Book V 'The Magnetic force is animate, or imitates a soul; in many respects it surpasses the human soul while that is united to an organic body'. A rhetorical *tour de force*, it began with the assault on Aristotle's great error of denying soul and motion to the Earth, and ended with him granting it some form of consciousness. He countered various objections – for example, animals have organs; that is how they are able to move.

So too, said Gilbert, did the Earth and other planets, 'albeit that these organs are not made up of viscera as animal organs are'. The Earth's magnetic poles and axis were evidence that it had the necessary organs for rotational movement. Gilbert's conviction that Earth's polarity or verticity had a physical, organic cause explains why he reminded readers that 'on the terrella the equinoctial circle, the meridians, the parallels [of latitude], the axis, the poles, are natural limits: similarly on the Earth these exist as natural and not merely mathematical limits'.

Gilbert believed that he had shown experimentally

how an animate form gave magnetic substance the
properties of poles, axis and rotational motion. The Earth
was a magnet. It was time to discuss the motion of the
Earth.

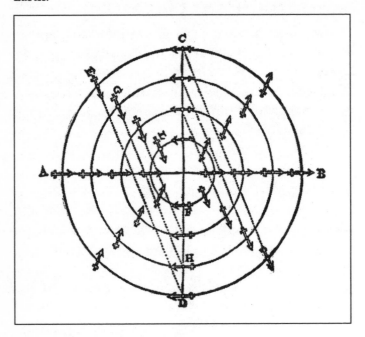

Illustration 10: 'The formal magnetic act
spherically effused'.
In this diagram, the solid terrella is represented by the
circle LF. The outer circles are parts of the sphere of virtue,
which extends infinitely. The inclining versoria with
dotted lines of inclination are all placed at latitudes of 45°,
as was the versorium in Ilustration 9. Notice how each
versorium is shown to incline to a point in its own sphere
(as G to H, or E to D), exactly as the versorium N pointed
to D in Illustration 9. The difference is that these points
(H, D) lie beyond the body of the terrella, suggesting that
the magnetic form is 'effused' and acts immaterially to
order 'space' as well as matter.

CHAPTER 13

GILBERT, COPERNICUS AND THE GIANT LODESTONE

We may find magnetic Copernicanism unconvincing, but many in the seventeenth century were persuaded. The modern discipline of astronomy was born in the decades after Gilbert's death. *De Magnete* accelerated the birth in a way that would have horrified him! But the fundamental transformation of astronomy was not the replacement of Ptolemy's geocentric universe with Copernicus' heliocentric one. It was the synthesis of a new science, which we call physical astronomy, out of two traditional disciplines, mathematical astronomy and natural philosophy. Astronomy was different, and inferior to philosophy. This attitude had to change before physical astronomy could take off. Some of those who changed it, like Tycho Brahe, vigorously objected to the Earth's motion. Gilbert was unique – a Copernican who did not want it to change!

The Astronomer's Circles and the Philosopher's Spheres

Gilbert learned the reason why astronomers were inferior at Cambridge. It was one scholastic doctrine he did not entirely reject. Astronomers dealt with planets whose orbits approached the ideal of uniform circular motion, but centuries of observations showed that they deviated from it in a number of small but annoying ways that philosophers could not explain. A few years after Gilbert's death, Johann Kepler made the radical claim that many of the problems disappeared if planets were allowed to accelerate and decelerate in slightly elliptical orbits, which we also believe.

Renaissance astronomers were given licence to come up with a variety of geometrical constructions, which they called theorics or hypotheses. These had the pragmatic aim of predicting accurately where the heavenly bodies would appear in the sky to an observer on Earth. Unlike the hypotheses of modern physical astronomers, Renaissance astronomers were not concerned whether their theorics were true or not. Ptolemy's own hypotheses used a geometrical device called an equant, which they knew was physically impossible.

Leaving detail and accuracy to the astronomers, sixteenth-century natural philosophers pontificated about a simpler universe. Their elevated debates concerned the physical causes of the basic heavenly motions. By 1600, only a handful of (primarily mathematical) scholars had converted to Copernicanism. Most philosophers were prompted to restate and develop old reasons for the Earth's immobility. Gilbert took them on as no one had done before.

The universe of traditional Aristotelian natural philosophers was a concentric nest of solid spheres, like an onion with the Earth at the centre. The heavenly forms gave the property of natural circular motion to the substance of their spheres. Closest to the Earth was the sphere (actually a combination of spheres) of the Moon, which carried the lunar body round in a monthly orbit. Then came the spheres of Mercury, Venus and the Sun, with orbits of a year or less. The sphere of Mars had a period of two years, Jupiter twelve and Saturn thirty.

Pressing down on Saturn was the 'sphere of the fixed stars'. This huge sphere rotated every twenty-four hours. As Copernicus pointed out, and Gilbert repeated, this rapid rotation messed up a harmonious relationship between distance from the centre and orbital period.

Beyond the stars that formed the visible boundary of the universe, Aristotelians located the sphere of the 'primum mobile'. Aristotle had proposed this prime mover because of his principle that everything that moved had to be moved by something else. To prevent an infinite regress of movers, Aristotle concluded that all the motions of the heavenly spheres originated from this entity, which itself was unmoved. It was a bit like God, and used by Christian Aristotelians as a proof of God's existence. The Prime Mover instilled motion in the sphere of the stars. The stellar sphere communicated it in turn to Saturn's. And so it passed it down to the Moon, even into the sublunary world, where it was a source of winds.

Below the boundary of the super- and sublunary world, of course, a different physics applied – the physics of the four elements that Gilbert detested. Down here, natural motions were rectilinear, not circular. They had beginnings and ends. Endless motion did not befit the Earthly realm of corruption. A fundamental Aristotelian principle was that things only moved 'for the sake of obtaining rest in their natural place'. The only natural motion of elemental earth was a rectilinear motion towards rest at the centre of the universe. Ignoring objects like a missile thrown upwards 'against its nature', all earthy things seemed to obey the law.

Can the Earth Move?

In Book VI of *De Magnete*, Gilbert rehearsed some of the arguments against a moving Earth, which he had found conveniently gathered in Copernicus' book *De Revolutionis Orbium Coelestium* [*On the Revolutions of the Heavenly Spheres*], published as Copernicus died in 1543. They had their foundations in the doctrine of the single, natural, rectilinear motions of the four elements. They were deployed against the 'semi-Copernican' position that the Earth only rotated diurnally, which was all that Gilbert defended in *De Magnete*. They applied equally to Copernicus' annual motion, though that raised yet more objections, especially among theologians.

If the Earth rotated on its axis then, even in the short time it took for a stone to drop from your hand to the ground, the Earth would have whizzed eastward. The stone would appear to fall in an arc to the west. In any case, if the Earth moved, you would surely be unable to stand. Clouds would race through the sky. Birds would never keep up with their perches when they took off. Buildings would collapse. Indeed, the Earth's entire solid sphere would disintegrate under the tremendous centrifugal forces.

Galileo later formulated something like our response. Terrestrial bodies are moved by two forces. The force of gravity pulls them downwards, while an inertial force impressed by the moving Earth makes them keep up with it. According to Aristotle, that gave Earth two natural motions, and broke the laws of nature. Copernicus had actually given his planet Earth three motions. Besides the diurnal rotation and annual orbit, he had to account for a phenomenon known as the 'precession of the equinoxes'. The phenomenon arises because the Earth's axis performs a tiny, slow gyration with a period of 26,000 years. Traditional astronomers and philosophers had seen the same changes in the heavens; they invented another sphere to explain it.

Before Gilbert, no Copernican had systematically reworked basic matter theory in order to answer the powerful Aristotelian objections. Copernicus had made a very unconvincing start. He was an innovative astronomer, but a traditional natural philosopher. He simply stated that if the Earth moved, then earthy things had to move in natural circles. As for things made of elemental water and air:

We would only say that not merely the earth and the watery element joined with it have this motion, but also no small part of the air and whatever is related in the same way to the earth. The reason may be either that the nearby air, mingling with earthy or watery matter, conforms to the same nature as the earth, or that the air's motion, acquired from the earth by proximity, shares without resistance in its unceasing rotation.

It was just as well that Copernicus primarily rested his case upon mathematical not physical arguments. He believed that heliocentric astronomy was more accurate, economical and harmonious. Gilbert repeated one such argument in Book VI. By letting the Earth rotate diurnally, Copernicus had stopped the motion of the stars and eliminated Aristotle's 'primum mobile', which disrupted the pleasing correlation between orbital radius and periodic time.

Copernicus was the first physical astronomer of modern times. In his preface to *On the Revolutions*, he expressed his disgust with the purely predictive role given to contemporary astronomy. He thought it the duty of astronomers to play a part in discovering the true way that God had created the cosmos. A big factor in his revolt was his exposure to Italian Neoplatonism, according to which the Creator had designed the cosmos using geometrical principles. Copernicus eliminated Ptolemy's equant device because it offended those principles. Heliocentrism was not a mere hypothesis, it was the physical truth. Mathematicians *could* determine natural philosophical questions. One of his proudest accomplishments was that heliocentrism allowed him to calculate for the first time the distances of all the planets from the Sun. Gilbert ignored it.

Unfortunately for the dying Copernicus, a colleague called Osiander inserted an anonymous foreword to *On the Revolutions*. It contradicted Copernicus' revolutionary vision of the disciplines. It advised readers not to be startled by the book's references to the Earth's motion. Everyone knew that:

> *since different hypotheses are sometimes offered for one and the same motion (for example, eccentricity and an epicycle for the sun's motion), the astronomer will take as his first choice that hypothesis which is the easiest to grasp. The philosopher will perhaps rather seek the semblance of the truth.*

Osiander's foreword encouraged people in Gilbert's era to treat Copernicanism as just another hypothesis. Some astronomers preferred it. It promised to be more accurate, the equant had been a pig of a calculating device, and Copernicus's treatment of technical problems like precession was truly impressive. They called Copernicus 'the Restorer of Astronomy', but they didn't think that the Earth moved. That was physically absurd. And they didn't argue about it, because physics was not their business.

The Rise of Real Copernicanism

Nevertheless, the first people to break ranks and argue publicly that the Earth moved were mathematicians like Copernicus. Men like the young Kepler and Galileo, Simon Stevin in Holland and Thomas Digges in England had the ability to understand his mathematical astronomy. They also shared his belief in the power of mathematics to discover physical truths.

For example, Digges was sure from his astronomical observations and calculations of the 'new star' of 1572 that it was above the sphere of the Moon, and that he had disproved the immutability of celestial matter. So was Tycho Brahe, a fine example of the many progressive astronomers who also thought the Earth's motion was physically absurd. Tycho and Digges also agreed that a comet of 1577 had moved through the heavens, which proved that there were no solid spheres. But if there were no solid spheres to move the planets, then what could move the Earth?

The small camp of public Copernicans had something else in common. They did not work in universities, where the subordination of mathematics to natural philosophy was maintained, and reflected in salaries and status. Just as Gilbert had pursued his anti-Aristotelian natural philosophy not in Cambridge but in the freer air of London, so the first Copernican mathematicians found positions as clients of European nobles and princes. So did Tycho. As with all Renaissance patronage, there was a trade-off. Patrons got a reputation for backing bright young stars. The astronomer clients were allowed to pass themselves off as natural philosophers!

There was one exception – William Gilbert. In his history of astronomy, Gilbert had praised mathematicians for being clever, hard-working people doing a difficult job. It was so difficult that they were forced to deal only in useful fictions. Every time natural philosophers had made their fictions real, as they had done with the heavenly spheres, disastrous errors had ensued. Gilbert believed that astronomers and natural philosophers should stick to their disciplinary roles. When Gilbert praised Copernicus on the very last page of *De Magnete* as the 'restorer of astronomy', he meant the traditional fictionalist astronomy of Osiander's preface.

Thus Gilbert was the only true Copernican who denied that Copernicus was a revolutionary astronomer. He did not like mathematical arguments for the Earth's motion. And, uniquely among the early Copernicans, he did not need them. He did not need to counter the unanswered arguments of Aristotelian physics with the mathematical superiority of heliocentrism. He had replaced Aristotelian theories of matter and motion with experimental magnetic philosophy. His magnetic philosophy of the Earth gave him all the proofs of the Earth's motion he needed and, as a natural philosopher, all that he would accept.

The Earth's Magnetic Motions

Gilbert had warned readers that *De Magnete* would climax with 'a consideration of the whole Earth; and here we decided to philosophise freely, as freely, as in the past, the Egyptians, Greeks and Latins published their dogmas'. He did not mean readers to treat his talk of the Earth's magnetic motion as free speculation – as some have. He meant the freedom from simplistic, crushing orthodoxies, which he believed had existed in the age before Aristotle. He expected orthodox philosophers and theologians to get upset, and he was right. But 'let the theologues reject and erase these old wives' stories of a so rapid revolution of the heavens which they have borrowed from certain shallow philosophers'.

In the climactic Book VI, he conceded that some heliocentric doctrines were only 'probable'. But 'we infer, not with mere probability, but with certainty, the diurnal rotations of the Earth'. Gilbert homed in on a stark choice. What rotated every twenty-four hours? Was it the whole sphere of the stars, and the even bigger sphere of the 'primum mobile' beyond that? Or was it the Earth? The choice was a natural philosophical one. Astronomers worked equally well with geocentric or heliocentric hypotheses. The crucial question was: which was better, Aristotelian or magnetic philosophy?

Gilbert, whose universe was infinite, said that it was impossible to prove that the stars were confined to a rotating sphere.

How far away from Earth are those remotest of the stars: they are beyond the reach of eye, or man's devices,

or man's thought. What an absurdity is this motion.
... Far more extravagant yet is the idea of the whirling
of the suppositious primum mobile.

Ignoring the Aristotelians' principle that the heavens are made of stuff with a natural circular motion, he turned their arguments on their head. The heavenly spheres were more likely than the Earth to disintegrate under centrifugal forces, even if they were made of iron [!], because they were bigger. And, besides the disharmony and 'tyranny' that a 'primum mobile' introduced to the heavens, Gilbert the empiricist asked for any evidence, even one tiny observation, that confirmed its existence. There was none. Contrast that with the magnetic Earth.

We therefore, having directed our enquiry toward a
cause that is manifest, sensible [that is, detected by
the senses], *and comprehended by all men, do know*
that the Earth rotates on its own poles, proved by many
magnetical demonstrations to exist. For not only in
virtue of its stability and fixed permanent position [of
the axis] *does the Earth possess poles and verticity ...*
By the wonderful wisdom of the Creator, therefore,
forces were implanted in the Earth, forces primarily
animate to the end that the globe might, with stead-
fastness, take direction, and that the poles might be
opposite, so that on them, as at the extremities of
an axis, the movement of diurnal rotation might
be performed.

In short, Gilbert had solutions for the two great physical problems of late sixteenth-century Copernicanism.

What force moved the Earth in circles? Magnetic coition. How, if there were no solid spheres, did the moving Earth remain stable, with its North Pole always pointing (roughly) to the Pole Star? Magnetic verticity.

There was one problem. It wasn't that Peregrinus' terrella failed to rotate. Gilbert 'doubt[ed] if there is such a movement'. The weak point was the precession of the equinoxes. Copernicus had explained it as a slight wobble of the Earth's axis. Did this mean that the Earth was not magnetically stabilised? Like any good scientist, Gilbert did not let a small anomaly upset a good theory. In fact, precession fitted neatly into the grand cosmology he described in *Nova Physiologia*.

The Earth's 'Astral, Magnetic Mind'

De Magnete affords some glimpses of the Gilbertian universe. It is a cosmos where all the stars and planets have souls, whose immaterial virtues extend infinitely until they interact. The planetary souls co-operate, inciting each other via a 'consentient compact' to move in mutual harmony for the good of all. The ratio linking a planet's distance from the Sun with its orbital period was a good example of that harmony.

It *was* the good of *all*. Gilbert deplored the literally geocentric way in which conventional natural philosophers and theologians thought only about how stars and planets affected the Earth. They were forced to invent an entire celestial mechanism just so the Earth could stand still. The Earth was not unique, and it certainly wasn't uniquely inert and corrupt. However, the 'Sun (chief inciter in nature), as he causes the planets to advance in their courses, so, too, doth bring about this revolution of the [Earth's] globe by sending forth the energies of his spheres [of virtue] – his light being effused'. Can there be any doubt that William Gilbert was a full Copernican?

The Earth's soul, indeed 'her astral magnetic mind', was part of the compact. She therefore acts to ensure the good of all. The daily rotation was necessary to spread the Sun's light over the whole Earth. Otherwise, half of the Earth would scorch, and 'could not give life to the animate creation, and man would perish. In other parts, all would be horror, and all things frozen stiff with intense cold'. Likewise, the Earth's axis is tilted with respect to the ecliptic by 23°. 'This has been wisely ordained by nature and settled by the Earth's primary eminency' because without it we would not have the seasons of the year and the fertility they bring.

Precession was another good design idea from the astral magnetic mind.

Thus do all the stars change their light rays at the Earth's surface, because of this magnetic inflection of the Earth's axis. Hence the ever new [dates of the] changes of the seasons; hence are regions more or less fruitful; hence changes in the character and manner of nations, in governments and laws, according to the power of the fixed stars.

Gilbert's profound belief in astrology was not unusual, but his belief that astrological forces were master-minded (mistress-minded?) by a terrestrial magnetic soul certainly was.

Buried in *Nova Physiologia* was a strange consequence of Gilbert's theory of the consentient compact of souls. Although the planets' motive forces were produced by souls, they were governed by laws that natural philosophers could investigate. Gilbert had shown how an effused sphere of magnetic virtue generated a law of inclination and latitude. He had also proposed an inverse law relating distance from a magnet with its power to attract.

With these conceptual resources, Gilbert could have tried to do with magnetism what Newton did with gravity. He could have tried to generate the exact orbits of the planets from laws of the forces acting on them. He went a little way down that road when he discussed the Moon, but he stopped. I think the reason was that he thought the dynamic interactions of the animate forces were just too complicated. Like the positions of the stars, the resultant motions were 'beyond man's devices, or man's thought'.

The idea that the celestial dynamics implied by magnetic philosophy were impossibly complex might also explain his attitude to astronomers. He wrote in *Nova Physiologia* that hypotheses . . .

> . . . were not the ravings of astronomers, even though they are different and contradictory, but all the learned men who investigate the phenomena are allowed to search for hypotheses which are more convenient for calculation. This must necessarily happen, even if the true Philosophy of bodies and science of the heavens were understood and well known.

Perhaps Gilbert was not a complete fuddy-duddy about

mathematicians after all. The inventor of the first plausible matter theory and dynamics for a Copernican universe pitied them because of the task left by magnetic philosophy.

De Magnete ends abruptly, after several pages of the technical astronomy of precession. 'All these points touching [precession] are undecided and undefined, and so we cannot assign with any certainty natural causes for the motion. Wherefore we here bring to an end and conclusion our arguments and experiments magnetical.' *Finis*.

That had been the aim of Book VI. Gilbert believed he had assigned with certainty natural causes for some motions of the Earth. It answers the puzzle of why he never discussed the Earth's annual motion. The yearly orbit was caused primarily by luminous solar virtue, not magnetic. You couldn't experiment on solar virtue. The terrella had been a marvellous model Earth. The voyages of the versorium had discovered the workings of the Earth's magnetic form. But the terrella was not a miniature Sun.

I conclude that we must take *De Magnete* seriously as a thoroughly experimental book. To be sure, we do not believe that an Earth soul or even diurnal rotation can be inferred from experiments with lodestones and navigational data. *De Magnete* went way beyond mere observations. Its forms of inference were methodologically naïve. Its experiments were constantly interpreted in the light of pre-existing theory and metaphysics, but what work of modern science does not do the same?

That conclusion does not help to solve a big mystery of Gilbert's work. If *De Magnete* was the first great success story of experimental science, why did Gilbert abandon experimentalism in *Nova Physiologia*?

· CHAPTER 14 ·

THE DARK SECRET OF
DE MAGNETE

Modern science is a collaborative enterprise, combining expertise beyond the competence of one person. Scientific papers signal their multiple authorship. Early modern science seems more heroic, the product of individual geniuses like William Harvey, Galileo, Gilbert himself. Of course, they owed debts to their education, forerunners, obscure colleagues, unsung technicians, but their books were their own work, weren't they? The dark secret of *De Magnete* is that Gilbert had a lot of help. Was the novel theory of the magnetic Earth his own? Undoubtedly. Was the novel synthesis of philosophy and experiment, theory and practice all his own work? Probably not.

Consider these conundrums. *De Magnete* is full of experimental and empirical evidence, while *Nova Physiologia* is not. *De Magnete* celebrates the practical application of theory. *Nova Physiologia* is unconcerned with it. *De Magnete* contains sophisticated geometrical representations of the magnetic force, yet Gilbert denied the usefulness of mathematics to natural philosophy. *De Magnete* and *Nova Physiologia* include some technical Copernican astronomy, when Gilbert was not an expert astronomer. *De Magnete* describes the most recent literature, practices and errors of navigators with an expert's precision, though Gilbert was no practical expert. *De Magnete* contains descriptions of new instruments and their use, but in a technical language he never used elsewhere.

When Marxist history of science was fashionable, a lot was made of Gilbert's declared and undeclared debts to navigators, instrument makers and miners. Gilbert got his experimental method from the sophisticated know-how of downtrodden craftsmen, it was said. The example of Robert Norman shows that there is something to this 'vulgar Marxist' interpretation.

In recent years, historians have focused on a subtly different group whom they call 'mathematical practitioners'. It includes enquiring instrument makers like Norman, but also decidedly non-'working class' gentlemen like Edward Wright. What united mathematical practitioners was their application of mathematical knowledge to 'practical arts' such as navigation, surveying, instrument making, even music making. The work of these systematic, problem-solving people led them to perform trials with lodestones, compasses, cannons, building materials, telescopes and so on, which can look very like modern experiments.

Marxists pointed out that the status of these men and their knowledge rose in Elizabethan England, when London was a booming centre of early modern capitalism. But capitalism hardly explains why the ruler of Florence wanted as his client a trouble-making Copernican astronomer called Galileo, who insisted upon the title of 'court philosopher'. There were many reasons why mathematicians were employed in European courts, and why they began to represent themselves as natural philosophers. After Gilbert's death, the integration with natural philosophy of 'high' mathematics such as astronomy, and 'low' mathematics such as navigation continued. Historians currently believe that this process was very important in the creation of modern disciplines such as physical astronomy, experimental and applied physics. Someone like Galileo, who revolutionised all three, combined the traditionally separate roles of mathematician and natural philosopher.

The Elizabethan court was not one where philosophically minded mathematicians found much encouragement. It did not match some Italian, French and German courts in patronage of natural philosophy. Elizabeth and her nobles were not rich by some European standards. They had threats of invasion and an emerging empire to worry about. They wanted mathematicians for pressing practical reasons. John Dee lost patience and emigrated. Thomas Digges gave up astronomical research for military work. Edward Wright found himself summoned from Cambridge to a ship bound for the Azores. *De Magnete* had no patron.

Edward Wright's De Magnete?

In *De Magnete*, Gilbert seemed to have created the experimental science of magnetism while keeping mathematics and natural philosophy separate. It may be significant, then, that *De Magnete* was in part a co-operative effort between Gilbert the natural philosopher and Edward Wright the mathematician. The accusation is that Wright wrote several parts of *De Magnete*, and not just his 'laudatory address' to Gilbert. Our chief witness is Dr Mark Ridley. Ridley was Gilbert's friend and experimented with Gilbert's own apparatus. He may even have been his lodger. He certainly knew Wright. After Gilbert's death, Ridley took on Gilbert's mantle of the radical magnetic philosopher, and when Gilbert's conservative colleague William Barlow criticised his magnetic Copernicanism, Ridley published a spirited defence. Ridley revealed that Wright had helped Gilbert, and Wright was a mean mathematician.

Ridley reminded Barlow that Wright had endorsed the magnetic motion of the Earth in his 'address'. Wright . . .

was a verie skilfull and painefull man in the Mathematickes, a worthy reader of that Lecture of Navigation for the East-India Company . . . [T]*his man took great paines in the correcting the printing of Doctor Gilberts booke, and was very conversant with him, and considering of that sixt booke* [of *De Magnete*] *which you* [Barlow] *no way beleeve, I asked him whether it was any way of his making or assistance, for that I knew him to be most perfect in Copernicus from his youth, and he denied that he gave any aide thereunto, I replied that the 12 chapter of the 4 Booke must needs be his, because of the table of the fixed Starres, so he confessed that he was the author of that chapter, and inquiring further whether he observed the Author* [Gilbert] *skilfull in Copernicus, he answered that he did not, then it was found that one Doctor Gissope* [Joseph Jessop] *was much esteemed by him, and lodged in his house whom we knew alwaies to be a great Scholler in the Mathematick, who was a long time entertained by Sir Charles Chandish, he was a great*

assistance in that matter as we judged, and I have
seen whole sheetes of this mans own hand writing of
Demonstrations to this purpose out of Copernicus, in a
book of Philosophie copied out in another hand[.]

These were extraordinary claims. Ridley, Wright and Barlow were three men with whom Gilbert discussed his work at length. All knew that the ex-mathematical examiner of St John's College, and England's most famous Copernican, couldn't understand Copernicus's mathematics. Gilbert's conservative attitude to mathematical astronomy makes that unsurprising. But Book VI of *De Magnete* gives the impression of someone at ease with Copernicus, even with complicated explanations of precession. Were those odd concluding pages Joseph Jessop's work? Perhaps technical astronomy got into *Nova Physiologia* when Gilbert's brother mistakenly included 'whole sheetes of this mans own hand writing'. We will never know, but a Gilbert stripped of competence in Copernican astronomy fits his self-image as a natural philosopher.

More questions arise. Which of the many competences combined in *De Magnete* did Gilbert actually have. Which chapters did he actually write? Ridley was surely right that Gilbert did not write the chapter on how to find variation.

Wright had published his *Certaine Errors of Navigation* the year before *De Magnete*. It included 'Tables of the Declination of the Sunne and fixed Starres', which reappeared in his chapter for *De Magnete* and betrayed him to Ridley. So did his design for an azimuth compass. In *Certaine Errors* he merely stated that he had devised an especially accurate instrument for measuring variation, which became known as his 'Mariners' rings'. There it was, described and illustrated in *De Magnete* IV, chapter XII.

When you start looking, Wright's expertise pops up in unexpected places. *Certaine Errors* also contained the first discussion in a navigation manual about the effects of atmospheric refraction. *De Magnete* duly noted that 'the experienced observer [of variation] will allow somewhat for refraction'. *De Magnete* has a throwaway remark about the 'island of St. Helena (whose longitude is less than it is usually given in maps)'. That was a major point in *Certaine Errors*, where Wright had published his new projection for maps in order to correct such errors. Wright probably volunteered directly his intimate knowledge of errors in compass construction and use, collations of variation data and so forth. Most likely it was he, with his contacts in the world of Dutch navigation, who alerted Gilbert to Stevin's *De Havenvinding*.

When you know that Wright definitely wrote Book IV, chapter XII, some technical and stylistic oddities contrast themselves with the bulk of *De Magnete*. There are italicised subheadings, specialist terms from practical mathematics, and mention of advanced techniques such as prosthaphaeresis, Tycho Brahe's precursor of logarithms.

Alert to these peculiarities, you become suspicious of other sections. The worst case is Book V, chapter VIII. It is the high point of applied mathematics in *De Magnete*, the extraordinary geometrical construction of the inclination–latitude relationship (see page 154). It is unlikely that Gilbert had the geometrical skill, patience and interest to produce and refine it. Wright, however, was a skilled mathematician, cartographer and navigation theorist who emphasised this practical application of magnetic philosophy in his 'address'. As well as the ingenious mathematics, the chapter has sub-headings and the same mathematical terms. It also mentions a new use of 'our variation instrument' – the same instrument of Wright's described in Book IV, chapter XII.

If Gilbert was not good at making, using or writing about navigational instruments, then who invented *De Magnete*'s impressive inclinometer? Wright's brother-in-law wrote to him in June 1600 about his 'new magnetic Instrument for the Declination', by which he meant decline or dip. When Wright added it to the 1610 edition of *Certaine Errors*, he gave the impression that it was his own. Let us grant that Gilbert himself first noticed the relationship between inclination and latitude when he experimented on terrellae. But Wright probably quantified the relationship, invented the nomograph and designed the inclinometer. If so, then Wright was responsible for the most novel navigational application of *De Magnete*, the magnetic finding of latitude. He may also have ensured that other applications, of variation to Stevin's longitude scheme for example, were fully explored.

Even if Wright's role was as large as this, Gilbert remains the man with the big ideas, the brilliant experiments, the bulky knowledge of lodestones – the butt-kicker of Aristotle, the creator of magnetic philosophy. The origins of his experimentalism remain controversial. However, the presence of a concealed hand in *De Magnete* (or two hands if we include Jessop the Copernican) does have implications. It suggests that Gilbert's use of the

Dip instrument

Illustration 11: Dip instrument.
De Magnete contained this illustration, and impressive instructions for the making and use of a fine dip instrument or inclinometer. Far from confirming Gilbert's expertise in practical mathematics and experiment, it was probably invented and described by Edward Wright. Wright wrote an introductory 'laudatory address' to Gilbert in *De Magnete*, but there is good evidence that he wrote other sections of the book, and collaborated with Gilbert to produce it.

'experimental' methods of craftsmen and practical mathematicians was less profound than we thought. It accounts for the contradictory role played by mathematics in *De Magnete*. And it leaves Gilbert looking more like a late Renaissance natural philosopher. People like that wrote books like *Nova Physiologia*.

De Magnete synthesised causal theory with quantitative, practical applications in a revolutionary way. It may be that the one book was the product of two minds and two approaches, still largely separate in many universities, in Elizabeth's court and in Gilbert's head. What were seventeenth-century readers going to make of it?

PART III

AFTER THE REVOLUTION: MAGNETIC PHILOSOPHERS INHERIT THE EARTH

Chapter 15

Goodbye Sailor! Gilbert's Earth Attacked and Defended

Gilbert's new magnetic philosophy took off almost immediately. *De Magnete* itself sold modest hundreds of copies, including pirate editions. But the doctrines spread through imitations, popularisations, summaries – and humdinging controversies. *De Magnete* engaged the interests of so many groups. Aristotelians were stung into refutations of its theory of matter. Conservative theologians and cosmologists were offended. Copernicans were excited. Occultists couldn't decide whether Gilbert's magnetic experiments were a challenge or an opportunity to make magic scientific. Mathematical practitioners generally embraced the new foundations of magnetic navigation. But one man with a longitude plan was not happy. The intensely practical interests of position-finding at sea created the first storm.

The Attack from France

In 1602, as Gilbert's life drew to a close, a French nobleman issued a challenge to Gilbert's priority, and to his major conclusions. The author was Guillaume de Nautonnier, Sieur de Castelfranc-sur-Lot, en Languedoc. De Nautonnier was a sailor by name, but not by vocation. He was a French oddity to match Denmark's Tycho Brahe. Like Tycho, he was an aristocrat who chose

Illustration 12: Title page of the second, pirate edition of *De Magnete* (Szczecin, 1628).

Unauthorised versions of marketable books were common. Lochman was a printer of mathematical works in a thriving port on the (modern) border of Poland and Germany, who issued more copies in 1633. The eye-catching images promise a book that combines natural philosophy and practical mathematics. Top left is the experimental terrella (cf. Illustration 6), while top right is a practical, capped the middle-class profession of mathematics. Like Tycho, he served his king as a mathematician: he was Geographer Royal. In a most extraordinary parallel, he constructed France's first astronomical observatory close by his chateau in 1609. The chateau and some remains of his observatory still stand, advertised as a tourist destination for amateur astronomers and ley-line lovers.

De Nautonnier published only one book, but it was huge in size and promise. *Mécometrie de l'eymant* [*The mecometry of the lodestone, or the way of determining longitudes by means of the lodestone*] described his solution to the longitude problem. It was a lavish work, written not for mariners but for monarchs and their mathematical advisors, with patronage in mind. It was also a white elephant.

De Nautonnier insisted that he had concluded independently of Gilbert that the Earth was magnetic. He might have done; the basic theory could have occurred to him, as to Gilbert, as a simple inversion of Peregrinus' ideas. But the good sieur produced no experimental or deductive proofs that he did not find in *De Magnete*. While de Nautonnier cited Gilbert's evidence approvingly, on fundamental principles they were literally poles apart.

oval lodestone. A philosopher holds a marvellously suspended chain of stones, faced by a sailor and the iron tools of his trade. Bottom left is the investigative inclinometer (cf. Illustration 11), opposed by the geometrical device for finding latitudes (cf. Illustration 9). At bottom centre, philosophy and navigation are united as a floating terrella and compass-guided ship. The title (cf. Illustration 5) emphasises that the work is a 'new natural philosophy'.

The *Mécometrie* brought to futile perfection a model that had developed throughout the sixteenth century, and that Gilbert's philosophy tried to destroy. De Nautonnier was the first person to propose explicitly that the Earth was a tilted dipole. Iberian magnetic longitude schemes had been consistent with the theory, but none actually mentioned poles. De Nautonnier did. He told his readers where they were. Their latitudes were 67°N and 67°S, just on the Arctic and Antarctic circles. Their longitudes (from our modern Greenwich meridian) were approximately 30°E and 150°W. His dipoles were thus located in Siberia and Antarctica. They were a considerable distance from the geographical poles of Gilbert's philosophy, though still in uncharted territory.

Magnetic Latitudes and Longitudes Galore

To establish his poles as a 'mecometric' or longitude-finding solution, de Nautonnier had to respond to *De Magnete*. In particular, he had to defend the sixteenth-century dream of regular variation from the attacks of Gilbert, Wright and Stevin. De Nautonnier's magnetic Earth was utterly spherical. Compass needles always aligned directly with magnetic North. There were magnetic meridians and parallels of latitude, but they were tilted or offset from the geographical grid by 23°.

The one great circle that ran through both pairs of poles formed two 'prime meridians' of zero variation. It ran through the Azores (30°E) and modern Indonesia (150°W), as had Portuguese meridians eighty years earlier. Sailing away from one of these meridians along

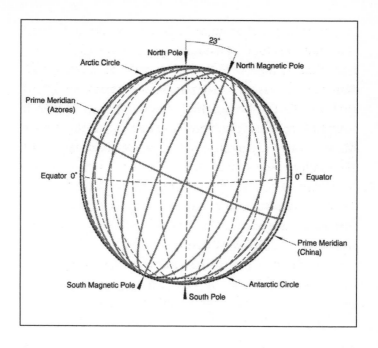

Illustration 13: De Nautonnier's mecometric meridians.
The diagram illustrates the longitude-finding potential of
de Nautonnier's 23° tilted dipole theory of geomagnetism,
which he first published in 1602. Two identical grids of
meridians and an equator have been superimposed on the
Earth. The dotted lines are normal geographical
meridians, converging on the North Pole and South Pole.
The heavy lines represent de Nautonnier's magnetic
meridians, converging directly at his magnetic poles. In
this elevation, the extreme edge of the Earth is the unique
great circle, composed of two prime meridians, that runs
through all four poles. The angle of intersection between
any two meridians gives the variation. At the
(geographical) equator variation would increase in both
directions from zero at the longitude of the prime
meridians to a maximum in the very centre of the
diagram. Such spherical geometry was the foundation of
most magnetic longitude schemes in the period.

the equator, variation would slowly increase to a maximum of 23° before decreasing towards zero again. Sailing north (or south) along any geographical meridian, variation would rise rapidly, to a maximum of 180° for points lying between the magnetic and geographical poles. Such huge predictions matched high latitude sailors' reports of big variations, but not their experience of irregular ones.

De Nautonnier's scheme required you to accept his axioms of separate poles and geomagnetic uniformity. After *De Magnete*, that was less likely. But if you did, then you saved yourself Stevin and Gilbert's project of carefully recording variation all over the world. All you needed was simple spherical geometry. From his Languedoc study, de Nautonnier saved people even that labour. The bulk of his book consisted of thousands of calculations, compiled into tables that de Nautonnier promised *nautoniers* they could use to read off their exact position from the magnetic needle.

Variation alone was not enough – many locations shared the same value. The complete mecometrician needed two other data. He needed to know which sector of the Earth he was in, as he ought to. Less trivially, he needed to know his latitude. *De Magnete*'s announcement of a regular inclination–latitude relationship, which de Nautonnier gratefully adapted for a tilted dipole, meant that magnetism really did seem to provide a complete position-finding method for any weather. The Sieur de Castelfranc expected a rush of royal envoys to his chateau.

They never came, but de Nautonnier's staggering challenge to Gilbert was not without foundation. It seemed to have been born refuted by the authoritative new achievements of Stevin the Dutchman and Gilbert the Englishman. Stevin had provided a careful empirical demonstration that variation data were irreducibly irregular, and Gilbert had added a theoretical demonstration that the irregularity was a consequence of the Earth's irregular magnetic geology. De Nautonnier had his answers.

The Case for Regular Variation

De Nautonnier's commitment to the Earth's sphericity led him to claim that Gilbert's theory of variation had a fatal flaw. Gilbert tried to have his cake and eat it. To explain variation, he had made the Earth a body marred by continental mountain ranges, ocean deeps and subterranean corruption. But when he argued that the Earth was like any other rotating planet, he dismissed these 'blemishes' as insignificantly small. De Nautonnier fastened on to the contradiction.

> *If someone says that the Earth's lack of roundness is the cause of the variation of the lodestone, I say that . . . it is absolutely certain and agreed that the Earth is round.* [Oceans, mountains and caves] *are almost nothing compared to the rest of the Earth, as is shown by William Gilbert in the last chapter of the first book* [of *De Magnete*].

Maybe, but not in Book IV. *De Magnete's* readers might have thought Gilbert's theory proved by the experiments using a terrella that 'crumbled away'. De Nautonnier did not try to replicate them. He was a traditional mathematician, not an experimental philosopher. In 1629 someone did – Niccolo Cabeo, a clever Italian Jesuit whom we will meet again defending Aristotle and Aquinas from Gilbert's attacks.

Seventeenth-century Jesuits were modern Aristotelians who were also good experimenters. Gilbert had not given the dimensions of his terrella with the model Atlantic Ocean, so Cabeo generously assumed an ocean depth of 20 miles. This is still only 0.25 per cent of the Earth's diameter. On a terrella 100 millimetres across, a model ocean is only 0.25 millimetres deep! Even when Cabeo improved upon Gilbert's design by filling his depression with wax, he triumphantly found that his terrella exhibited no detectable variation. Of course, Cabeo had not taken account of Gilbert's rider that the Earth's surface scars masked bigger subterranean damage.

Nevertheless, de Nautonnier had grounds for believing that Gilbert had not secured the hypothesis of irregular variation. As his immense labours show, de Nautonnier was convinced that the data supported him. He admitted that many variation data did not converge on his poles but, like Stevin, Wright and all other experts, he emphasised the unreliability of the innumerable observations of the compass needle made by 'people who do not know how to be guided by it'. He had rejected, he said, all but 'the observations of the greatest pilots who have lived in the world in the past age', which he estimated were 99.9 per cent trustworthy. Granted, he gave no good criteria for his selections, and was obviously guided by his conviction of magnetic regularity. Nevertheless, he asserted that his longitude scheme stood or fell on its empirical adequacy, and he issued the usual appeal for more observations.

De Nautonnier's theory was much more vulnerable to disproof than Gilbert's. In principle, all it needed was two reliable observations that did not converge at his poles. By contrast, the Stevin–Gilbert model was virtually irrefutable, because it predicted unpredictable patterns of variation. It offered almost no criteria for identifying rogue or suspect values, and could incorporate almost any result as further confirmation.

In modern geomagnetic theory both de Nautonnier and Gilbert were equally wrong – or equally right. De Nautonnier's conviction that the Earth had separate magnetic and geographical poles has been borne out. Moreover, the geomagnetic field is more regular than Stevin and Gilbert would have conceded. Some of their 'reliable' data must have been real rogues. Rogues are rejected by computer models of the geodynamo, which can be set to imitate the magnetic field in 1600 according to modern physics. On the other hand, the same models incorporate Gilbert's intuition that the Earth's complex internal structure is a significant factor.

What neither Gilbert nor de Nautonnier foresaw was the discovery in London, in 1634, that magnetic variation changes over time! So, as we now know, neither regularity nor irregularity theorists could have established their case unless they knew that all the observations they mobilised were reliable ones made in the same short period of time. The precise nature of the magnetic Earth was still up for grabs. So too were the prizes for the longitude.

French and English Reactions to de Nautonnier

If the English needed a reason to close ranks and protect their countryman's reputation as the founder of true magnetic philosophy and navigation, then de Nautonnier provided it. De Nautonnier was always more credited in France than in England. A royal mapmaker used his magnetic co-ordinates. But even in France the responses were mixed. When Henri IV's reign ended with his assassination in 1610, de Nautonnier suffered scientific assassination at the hands of the court mathematician Jacques Alleaume and the 'Professor of divine mathematics' Didier Dounot de Barleduc. Dounot's *Confutation de l'invention des longitudes* [*Confutation of the discovery of longitude, or The Mecometry of the Magnet*] took the obvious geometrical route of showing that many observations did not converge to a point or pole, and managed to find calculational errors. Dounot declared de Nautonnier guilty of impudence, ignorance and chicanery.

Despite Dounot's demolition job, de Nautonnier's edifice was still standing in 1625, when Marin Mersenne published his *La Vérité des Sciences* [*Truth of the Sciences*]. Mersenne was the correspondent and confidant of every European scientist. He offered de Nautonnier's approach as an example of his vision of scientific progress based on a careful synthesis of mathematics and observation. But as the 1630s closed, the incredible English discovery of secular variation began to be credited, even in France. De Nautonnier's scheme faded away.

Back in England, Gilbert's challenger had never been given house room. De Nautonnier's magnetic longitudes and latitudes were met with horror by London experts in navigation and magnetism. Their experience of variation, combined with Gilbert's and Wright's support for Stevin's empirical approach, led them to worry about misguided research and, worse, misguided mariners. Moreover, de Nautonnier attacked their countryman, Gilbert. Their nightmare came true when an Englishman published a suspiciously similar magnetic longitude scheme.

The Reverend Antony Linton, a little known Sussex parson, wrote his *Newes of the Complement of the Art of Navigation* in 1608. Linton only revealed the *Newes*, and not the precise position of the poles or the passage to China, the existence of which he claimed to have deduced. Linton was serving God and Mammon. He would only 'offer and tender unto you, and every one of you, my selfe and my services, upon meete and reasonable conditions'.

Linton dared to criticise Wright for lacking 'a good foundation whereon to raise his works, that is, the knowledge of the magnetical poles hereafter mentioned'. He knew that Gilbert, 'a great learned man and his followers, absolutely denie that there is any fixed pole magneticall, yet neverthelesse there are two'. The French disease had infected Gilbert's homeland, and his Jacobean successors mounted a campaign to stop de Nautonnier and Linton. Through the campaign we can see Gilbert's elevation to a national scientific hero, and his magnetic philosophy a legacy to be protected, promoted and fought over.

In the Court of the Philosopher Prince

Gilbert's coronation as England's king of natural philosophy occurred some time after James I had been crowned

Elizabeth's successor. Intellectuals benefited from James's image of himself as a bountiful philosopher–king. He was not 'the wisest fool in Christendom'. He chaired theological debates, patronised scholars like Isaac Casaubon, supported the eccentric philosopher of occult magic (and Gilbert enthusiast) Robert Fludd, gave jobs to Paracelsian physicians against the wishes of the College of Physicians, and tried unsuccessfully to entice the Protestant Copernican Johann Kepler away from the Holy Roman Imperial court.

It was in the satellite court of James's son, Prince Henry, that Gilbert's theories were developed and de Nautonnier repulsed. Gilbert's closest collaborators in magnetic philosophy got positions in Henry's court. Edward Wright stopped struggling as a navigation lecturer and watercourse surveyor, and served as a tutor and keeper of the library. William Barlow became the Prince's chaplain. Gilbert's half-brother, styling himself the Prince's 'most addicted client', jumped on the bandwagon and dedicated *De Mundo* to him. Wright and Barlow seized the opportunity to promote their versions of magnetic philosophy although, even in the new regime, their primary value was as experts in navigation.

Henry's court recalled the magnificence of Henry VIII with its patronage of learning, art, architecture and the sciences. His minders wanted to fashion him into a sophisticated, learned, Protestant prince, poised to lead England on to the imperial world stage. They built up art collections, a library of thousands of volumes, and fostered the boy's love of ships, sailing and the art of navigation. They backed programmes to extend English naval power, to recolonise America and to find the elusive

North-West or North-Eastern passage to the riches of the Orient.

The English fleet was employed in a Gilbertian research programme. Its explorers were instructed in Gilbert's theory of variation, which predicted an eastward passage. However, in his ill-fated 1610 voyage backed by the prince's court, Henry Hudson switched from north-east to north-west. He died when he was left by mutineers in the great Canadian bay named after him. A return expedition by William Baffin (of Baffin Bay fame) did more research. Baffin was baffled because 'the variation of the compass is 56 degrees West, which may make questionable Dr. Gilbert's rule, that where more Earth is, more attraction of the compass happeneth by variation towards it'. However, English navigators to the Cape of Good Hope confirmed it.

Linton and de Nautonnier threatened to disrupt the consensus. We can imagine anxious conferences of Henry's experts in magnetic philosophy, courtly backers of voyages and the leading navigators who came to St James' around 1609. One by one, English magneticians came out decisively in support of Gilbert.

Barlow refuted de Nautonnier in a manuscript of 1609, but first into print was Edward Wright in 1610 with a new edition of his *Certaine Errors*, now dedicated to Henry. A new error was being propagated by those 'of the opinion that there be two Magneticall poles, by knowledge whereof ... they have imagined that they could find the longitude'. Wright's refutation was predictably empirical. He published a table of observations that did not converge. Discrepancies were grist to his mill. He gave values for the vicinity of St Helena ranging from 3°W to 7°W. He warned that 'if any shall think that the great difference that is found betwixt divers of these observations taken at the same place by divers observers, doth make anything against the intention of this argument, he is much deceived'.

Wright's new edition made most of the novelty of inclination, the triumph noticed by 'our countriman Robert Norman, and then much more discovered by Dr Gilbert and that Reverend Divine your H[ighness's] Chaplaine Mr William Barlowe'. It afforded 'the like hope of finding as much in effect as the latitude, as the variation does of the longitude'.

Once again, the English fleet participated in research inspired by *De Magnete*. Wright reported that the magnetic method for latitudes had been 'approved and confirmed, by the observation and experience of divers of our best navigators at Sea'. Apart from a report by Jesuits sailing to China fifty years later, this is the only evidence we have of marine trials.

Learned magneticians constantly urged its value, especially in overcast conditions. In England, inclinometers were designed by Barlow and, of course, Wright. Practical tables derived from Wright's nomograph were widely printed. Gilbert himself authorised one in 1602. But a search of navigators' logbooks has found no evidence of actual observations. For all the English enthusiasm, the method was not really beneficial. The angle of inclination altered too slowly with latitude to be discriminating, especially given the difficulty of accurate measurement at sea. Although Gilbert believed that inclination, though not variation, was little affected by geological irregularities, modern geomagnetism has discovered many variables. Sailors did well to wait for the skies to clear.

In 1612 Henry, now Prince of Wales, died aged only eighteen. So successful had his courtiers been in fashioning his image that Britain was united in grief at the loss of the great helmsman apparent, and at the prospect of his younger brother Charles becoming king. The prestige of Gilbert's magnetic philosophy in Henry's court provided a motif for a commemorative sermon, published in 1613 as *Prince Henry his first anniversary*. The theologian Daniel Price recalled how:

> *a glimmering light of the Golden times appeared, all lines of expectation met in this Center, all spirits of vertue, scattered into others were extracted into him ... His Magnetique vertue drewe all the eies, and hearts, of the Protestant world.*

Bacon and Gilbert

One courtier who might have winced at these words was Francis Bacon, Lord Verulam, Viscount St Albans and Lord Chancellor to James I. Bacon, the consummate politician, lawyer and loyal supporter of James' aim to increase monarchical power, is better known today as the reformer of scientific method. His first major statement, *The Advancement of Learning*, appeared two years after Gilbert's death in 1605.

This book has mentioned many of the similarities between Gilbert and Bacon. Both were radicals who believed that natural philosophy needed a new start; both saw Aristotelian natural philosophy as the big

Illustration 14: Portrait of Francis Bacon (1561–1626). Francis Bacon, Viscount St Albans, had very similar views to Gilbert about the need for a new, experimental science, although he criticised Gilbert for making a philosophy out of the lodestone. He is depicted here at the end of his life, after his fall from political office, composing one of the six volumes of his *Instauratio Magna* [*The Great Instauration*], his blueprint for scientific reform. Bacon was a great writer *on* science, but not a doer of it like Gilbert (cf. Illustration 1).

enemy; both gave historical explanations of how it had become institutionalised. They had in common a post-Renaissance ideology of progress, the centrality of experiment in their vision of future science, and a preference for the practical natural knowledge of craftsmen over the meaningless abstractions of philosophers. Some have even described Gilbert as 'one of the first to vindicate the Baconian method', which certainly strains the chronology!

Bacon certainly didn't think so. Although he offered the occasional grudging word of praise, and used one or two magnetic facts gleaned from Gilbert's writings, Bacon's prevailing judgement was that Gilbert blew it. It was good that Gilbert had busied himself with experiments but, precisely because he lacked Bacon's method, he had been condemned to make the same mistakes as Aristotle, Paracelsus and all the other ancient and modern system builders. As Bacon put it,

> *The race of chemists* [that is, Paracelsians] *again out of a few experiments have built up a fantastic philosophy, framed with reference to a few things; and Gilbert also, after he had employed himself most laboriously in the study and observation of the loadstone, proceeded at once to construct an entire system in accordance with his favourite subject.*

This could mean that Bacon was good on method, but a poor judge of science. It is true that Bacon also rejected Copernicanism, Harvey's theory of the circulation of the blood and the general usefulness of mathematics, although in all of these judgements he was in good company.

Or did he have a point? Bacon probably knew as much about Gilbert's philosophy as anyone except Wright, Ridley and Barlow. He was one of the few to use and quote from *De Mundo* as well as *De Magnete*. When Bacon said that Gilbert made a philosophy out of the lodestone, he was right. Gilbert got carried away, but Bacon did not stop him carrying many seventeenth-century scientists with him. Until 1650 the question was not 'Is magnetic philosophy right?', but 'What kind of magnetic philosophy is right?' The battle for Gilbert's inheritance created fierce intellectual turf wars, beginning in London, and with two of his closest colleagues.

CHAPTER 16

HUMOURS UNRECONCILED: MAGNETIC WAR IN JACOBEAN ENGLAND

The royal centre of magnetic research was now dead. Its experts found themselves drawn by necessity to the scattered opportunities offered by London's commercial community. Wright must have feared a return to surveying, for his pay-off was small: 'Mr Wright the Keeper of the librarie, an excellent Mathematitian, & Navigator, & a very poor man – 30li'. By 1615 he was dead too. Barlow retained a church position, but he no longer had the credibility of being *de facto* a royal magnetician. The mantle of being Gilbert's English guardian would have to be earned among London's élite. Barlow was furious when a book called *A Short Treatise of Magneticall Bodies and Motions* appeared in 1613.

Mark Ridley, Magnetic Copernican

The author was Dr Mark Ridley. Ridley was Gilbert reincarnated – and fourteen years his junior. Ridley's medical career had followed Gilbert's from Cambridge to high office in the College of Physicians. He knew mariners too, having spent a few years in Russia as a physician to the resident English merchants and even the Czar himself. After he returned to London in 1598, he became Gilbert's friend and protégé in magnetical

A
BREIFE DIS-
COVERY OF THE
IDLE ANIMADVER-
SIONS OF *MARKE RIDLEY*
Doctor in P H I S I C K E vpon
a Treatife entituled,

Magneticall Aduertifements.

——————— *moueat Cornicula rifum.*
Furtiuis nudata coloribus ———

LONDON,
Printed by *Edward Griffin* for *Timothy Barlow*,
at the figne of *Time* in Paules Church-
yard. 1 6 1 8.

Illustration 15: Title page of William Barlow's
Breife Discovery (London, 1618).
Barlow and Mark Ridley engaged in a battle to don
Gilbert's mantle as the prime magnetic philosopher in
England. This bitter pamphlet was the last in a series of
exchanges provoked by Barlow's accusation of plagiarism.
Barlow mentions his own book, *Magneticall Advertisements*
(1616), but manages not to advertise his rival's! Barlow
wanted magnetic philosophy to avoid irreligious
Copernicanism and to concentrate on sound navigational
applications. Ridley extended Gilbert's terrestrial
magnetic form to all the planets.

philosophy. He said he had experimented using Gilbert's own apparatus.

If Gilbert and Wright had combined practical and philosophical interests as they finalised *De Magnete*, Ridley and Barlow showed how easy it was to pull them apart. Unlike Barlow, Ridley was not especially interested in magnetic navigation. He was like his mentor, more excited by the radical, cosmological, anti-Aristotelian implications of magnetic experiments. Ridley and Barlow locked horns in a nasty pamphlet war that made them famous. When the court playwright Ben Jonson used Gilbert's magnetism as the metaphorical conceit of his play *The Magnetick Lady, or Humours Reconciled*, he dropped their names.

Magneticall Bodies made its natural philosophical ambitions clear from the start. Ridley defined a 'magneticall body'. It was one 'which seated in the aether or air, doth remain and place itself in one kind of place or situation natural, not alterable'. Not exactly arresting, until Ridley gave examples: all the stars, planets and the satellites of Jupiter and Saturn.

Ridley's startling reference to satellites shows how far cosmology had come in the ten years since Gilbert's death. Galileo had trained a telescope on Jupiter in 1609, and published his discovery of four satellites in 1610, together with telescopic proof that Venus orbited the Sun. In 1612 he announced that Saturn had two companions (which we know as Saturn's rings). Ridley also cited Johann Kepler's confirmation of Galileo's discovery of sunspots, together with Kepler's inference that the Sun and other heavenly bodies rotated on their axes. *Magneticall Bodies* was the first English book to support these two famous Copernicans. Ridley's radicalism was unmistakeable.

Ridley's description of a completely magnetic cosmos was his novel contribution to magnetic philosophy. 'It is the vertue polar and Magneticall that holdeth all the globes in their position whatsoever.' Ridley also reasoned that, since a versorium completed two full revolutions when it made a full revolution of a terrella, the Earth's axis comprised not one but two polar pairs. And, of course, he lacerated de Nautonnier and Linton with a sharp wit that would have pleased his mentor. For the rest, *Magneticall Bodies and Motions* was *De Magnete* in English. Ridley declared that Gilbert's 'labours are the greatest, and best in Magneticall Philosophy'. He had staked his claim to succeed his 'friend and collegiat' as England's custodian of magnetic philosophy.

William Barlow, Divine and Dulbert

Barlow was not happy. In twelve months, he had lost his patron and priority in publication. It irked the conservative cleric that his pre-emptor was a philosophical radical, freely promoting a turning Earth, telescopes and tilted magnetic planets. But what really riled him was his conviction that Ridley had plagiarised his 1609 manuscript.

Barlow worked up his manuscript into a book called *Magneticall Advertisements*, published in 1616. Barlow's target audience was similar to Ridley's, the 'many of our Nation, both Gentlemen and others of excellent witts' who could not read Latin but who would nevertheless find, in the words of his subtitle, *divers pertinent observations, and approved experiments concerning the nature and the properties of the Loadstone ... very pleasant for knowledge, and most needful for practise, of travelling, or framing of instruments fit for Travellers both by Sea and Land.*

Barlow strove to establish his credentials and originality as a magnetic philosopher. Without naming Ridley, he referred to his 'propositions set abroad in another man's name, and yet some of them not rightly understood by the partie usurping them'. But how could he step out of Gilbert's shadow, especially now that Ridley had presented himself as Gilbert's heir? Barlow paid due reverence to Gilbert. He admitted that he had not believed his theory of the magnetic Earth until he had replicated the experiments and talked to navigators. But equally, he insisted that he had been researching magnetism since 1576, and that Gilbert had appreciated his contributions. To prove it, he included Gilbert's letter to him of 1602, thereby preserving our sole example of Gilbert's correspondence. Gilbert wrote that 'you have shewed me more – and brought more to light than any man hath done', but he condescended to him, offering 'to commend you to my L[ord] of Effingham', and agreeing to publish some experiments so that 'you may be knowen for an augmenter of that art'.

Barlow needed to differentiate his magnetic philosophy from Ridley's super-Gilbertian version. It wasn't difficult because, as it turned out, he had always disliked Gilbert's Copernicanism and his virulent attacks on Aristotle and ancient learning. The *Magneticall Advertisements* would support Books I–V of *De Magnete*, but not Book VI, which Barlow's love of truth compelled him to disown. That prompted Ridley to spill the beans on Gilbert's weakness in mathematics.

Magneticall Advertisements differs so much from Ridley's *Magneticall Bodies and Motions* that Barlow's charge of plagiarism is hard to credit, even if Ridley had seen his manuscript by 1613. In effect, both plagiarised Gilbert, albeit with different agendas. Barlow's book avoids cosmology, except to deny that the Earth moves. He trampled on Gilbert's most cherished principle when he marvelled at how God had put the navigationally useful property of magnetism into 'a base, contemptible and dead creature, as [the Earth] seemeth to be'.

Since Barlow rejected Gilbert's cosmology, he was not preoccupied with spherical magnets. Gilbert had been right to call a spherical magnet a terrella, because it was 'an exceedingly small model (as it were)', but Barlow concentrated on experiments with different, more practical shapes. He thought oval lodestones were best, because their stronger and more obvious poles made them more reliable magnetisers of compass needles. His unique selling point was a 'double-capped loadstone', made by his Winchester workman and mentioned in Gilbert's letter. *Magneticall Advertisements* went on to practical discussions of compass manufacture, instrument design and magnetic navigation.

Riled by the charge of plagiarism, Ridley rushed out his *Magneticall Animadversions. Made by Dr Mark Ridley, Doctor in Physicke. Upon certain Magneticall Advertisements, lately published, From Maister William Barlow*. In the way of pamphlet wars, his title aimed at wit: 'Animadversions' (that is, critical reflections) punned on 'Advertisements', while alluding to their dispute about whether the Earth was animated by magnetism. Ridley denied that he had stolen any ideas. The West Country clergyman knew no more than London workmen – Barlow had no ideas to steal.

'There is almost no proposition in this book which most Mariners[,] Instrument-Makers, Compasse-makers, Clocke-makers, and Cutlers of the better and more understanding sort around London and the Suburbs have not known, practized and made long before.' His so-called inventions were 'most of them in the Doctor Gilbert's Booke, as I said before, or else such ordinary things that any ingenious workman hath or may easily invent or make; unless you hold all men Dulberts like your rare workman of Winchester'.

Barlow responded with *A Breife Discovery of the idle animadversions of Marke Ridley Doctor in Phisicke*, which, as I write, is the only work of magnetic philosophy besides *De Magnete* stocked by Amazon.com! This time it was personal. He tried to discredit Ridley, suggesting that he had morally compromised himself in order to 'in so short a time become [the Russian] Emperors principall Physition'. In a *double entendre* to Ridley's observations or 'looks' with the new-fangled telescope, he insinuated that the youthful Ridley had seduced the Czar, 'for his lookes . . . are his meanes'.

Magnetic Copernicanism Under Fire

Barlow was also provoked to make the first public assault on Copernicanism by an Anglican theologian. England was not Catholic Italy and Copernicanism was not a crime of heresy. Nevertheless, cosmological radicals in supposedly tolerant countries could feel intense pressures to conform. In *Nova Physiologia* Gilbert had expected 'the theologians to cry out that I am insane and foolish'. The response to Ridley of Barlow, who had been the Prince of Wales's chaplain, gives an insight into English attitudes.

Since Ridley had challenged him to refute Book VI, Barlow let him have it, beginning with another personal attack:

> *Out of all question, somewhat it is more than ordinarie, that maketh him* [Ridley] *of so hauty a spirit, so to brave the world with such prodigious assertions of his Magneticals, in, and above the Moone, and his paltry abusing of the holy scriptures to support his lunaticke fictions under the name of Magneticall Philosophie.*

Barlow restated three well-known conservative rules, one each from theology, natural philosophy and mathematics. Biblical statements of the Earth's immobility are literally true; science is based on sense experience, and our senses confirm that the Earth stands still; astronomers like Copernicus only suppose 'the motion of the earth, for an Hypothesis, serving their ready calculations'. Barlow concluded that no educated person would believe Ridley's bad arguments. They 'may goe current in a mechanicall Trades-mans shop, yet they are very insufficient to bee allowed for good by, men of learning, and Christians by profession'.

Ridley, then, was no Christian gentleman, and Barlow invited him to look 'in his owne conscience (if he have any)' and admit plagiarism. Moreover, Ridley was irrational and mad – a literal 'lunaticke'. Ridley's loss of reason accounted for his attacks on Aristotle.

Barlow was no uncritical scholastic. Like Gilbert and Ridley, Barlow thought that experimental knowledge had the power to end the empty disputes of 'clerkes'. But he certainly preferred Aristotelian philosophy to that of Gilbert and Ridley. Forgetting that Aristotle had denied creation, and the immortality of souls, he said it was Ridley's 'guise to make contrary conclusions to that which the Scriptures affirme, and therefore blame him not, if he doe so despise Aristotle, who never taught any such Logicke'.

The Barlow–Ridley affair confirms that theories of a moving Earth could be discussed freely in England. True, the vast majority of the educated élite, especially professors and divines, thought that they were theologically offensive, scientifically unsound, even posturing madness. In a society governed by patronage, Copernicanism was a bad career move for anyone but the most protected or indifferent, but it could be discussed. However, to carry the fight to orthodoxy as Ridley had was intolerable. The consequence was not a prison cell or censorship, but character assassination. Others, like the mathematician and astronomer Thomas Harriot, chose not to flaunt their opinions.

Lost amid the rhetorical sound and fury, Barlow made a penetrating argument against Gilbert's magnetic philosophy. Ridley had defined a magnetical body, but what was a magnetical *motion*? Barlow seized upon the mutuality that underwrote Gilbert's concept of coition. It 'is a natural inclination of two Magnets or Magneticall bodies that may freely move'. But two were a pair. Barlow insisted that a single magnet was 'utterly voide of all intrinsecall or selfe-Motion'. It followed that magnetic Earth cannot be the cause of its own rotation. Since Ridley had defined all the rotating heavenly bodies as magnetic ones, Barlow concluded that Ridley could be ignored as a serious contributor to magnetic philosophy. He needed to 'confine his magnetismes to the Earth'.

Barlow's objection struck at the heart of Gilbert's philosophy – the central principle of analogy between terra and terrella. The terrella and all its phenomena were necessarily observed within the Earth's own magnetic sphere of virtue. The behaviour of a single terrella could never be observed unless, as someone observed drily in 1663, it was taken into outer space. Would it rotate diurnally or stand still? Would a decisive experiment ever be devised? These would become preoccupations of later magnetic philosophers.

Barlow thought that all magnetic cosmology was misguided speculation. We might be tempted to think that boring Barlow, not revolutionary Ridley, was the credible custodian of Gilbert's legacy. He had separated Gilbert's facts from his speculation. But science isn't like that. Some of the greatest minds of the seventeenth century were convinced by Gilbert that magnetism had to have cosmic significance. As the Copernican debate heated to boiling point with the trial of Galileo in 1633, Gilbert's legacy became even more disputed. His terrella–terra analogy was tenaciously pushed to breaking point by Copernicans and anti-Copernicans alike. Barlow's cautionary voice was not heard again for thirty years.

Chapter 17

Kepler, Galileo and the Magnetic Earth in Motion

Magnetic Copernicanism was a hot topic for more than fifty years. Two natural philosophers played big parts in closing down Gilbert's magnetic cosmology. One is very famous. In his *Principia Philosophiae* [*Principles of Philosophy*] of 1644, René Descartes showed that it was just possible to explain magnetism in material terms. He proposed that magnets, including the Earth, emitted screw-threaded particles. The threads ran one way for north poles and the opposite way for south poles. Attraction, or coition, occurred when the screws burrowed their way into the grooved channels that all ferromagnets possessed. I personally find Gilbert's immaterial form more plausible, and modern science has abandoned (once again) the idea that magnets emit something material. Still, mechanistic explanations like Descartes' banished the unease that many natural philosophers had with Gilbert's talk of soul.

The other is virtually unknown. His name was Jacques Grandami, a Frenchman who taught at Descartes' old college. Grandami's book appeared one year after Descartes'. One of its attractions was a magnetic longitude scheme, but the title made clear its main purpose: *Nova Demonstratio Immobilitatis Terrae* [*A New Demonstration of the Earth's Immobility, Sought from Magnetic Principles*]. He was convinced that it would settle the question that had long gripped *De Magnete*'s readers: had Gilbert proved that magnetism moved the Earth?

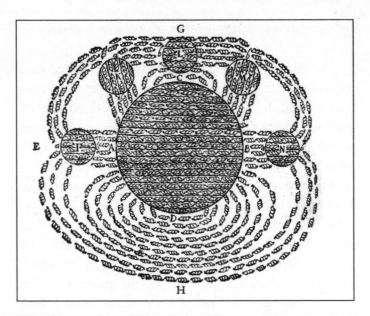

Illustration 16: Descartes' mechanistic explanation
of magnetism.
This diagram appeared in his *Principles of Philosophy* of
1644. The basic cause of magnetism was unique,
screw-threaded particles, not a form or soul. Left-handed
and right-handed threads accounted for bipolarity.
Attraction was caused when rotating particles screwed
their way into the suitably grooved channels that defined
a magnetic body. Because the particles circulated and
re-entered a magnet, there was no net emission or
detectable loss of weight. Note that Descartes' model
mechanical magnet is still Gilbert's terrella. He explained
geomagnetism using the same principles writ large.

Grandami was a rising star in the Society of Jesus, the
fiercely intellectual order that had secured Galileo's
humiliation in the trial of 1633. Provoked by Galileo,
Catholic cardinals had formally condemned theories of a
moving Earth as heresy in 1616. Like all Jesuits, Grandami

had been trained to defend this and other articles of Counter-Reformation faith. As tensions increased with Galileo's trial, philosophers like Grandami decided that magnetic arguments for Copernicanism had to be met head on and destroyed. The Jesuits' attacks on Gilbert were the culmination of an extraordinary fifty-year period during which not just the Earth but the whole universe became magnetic. Grandami and his brothers did not win the war, but they did manage a historic stalemate.

Not until the appearance of Isaac Newton's *Principia* in 1687 was there a workable physical explanation of how the Earth could career through space in orbit around the Sun. Philosophers and astronomers did not wait for Newton, however. Although Gilbert and Wright had been avant-garde in 1600, Copernicanism was almost an orthodoxy by 1687. Newton's achievement was to synthesise the achievements of Copernicus, Kepler and Galileo into a universal theory of mechanics. His key concept was that of universal gravitation. But Gilbert's immediate successors were trying to push the concept of universal magnetism.

Even before Gilbert's death, readers were hailing *De Magnete* as supplying the physics that Copernicanism lacked. Edward Wright, of course, had given his support in his 'address'. The next endorsement came from Paris in 1601. The book called *Philosophia Epicurea*, an eclectic and obscure mixture of atomism, animism and magnetism, was published by Nicholas Hill, an English courtier in exile who soon took his own life. Hill and Ridley were the only true Gilbertians, for the philosophical tide was already turning against ideas of planetary souls. But that left plenty in *De Magnete* for leading Copernicans to get excited about. By 1602, Kepler and Galileo were devotees.

So too was Simon Stevin in Holland. Doubtless it was the navigational applications that made him one of Gilbert's first readers, but Stevin was also one of the first Copernicans. Just as Gilbert had endorsed his longitude scheme, Stevin came out in favour of magnetic philosophy. Like most Copernicans except Gilbert, Stevin enthusiastically combined mathematical with natural philosophical interests. His book on *De Hemelloop* [*The Heavenly Motions*] contained the first full-blown statement of magnetic Copernicanism when it appeared in 1608.

Stevin shared much of Copernicus' conservatism. Unlike Gilbert, he would not abandon solid heavenly spheres, leaving him free to deny that the Earth moved magnetically. But he seized upon Gilbert's magnetic explanation of the fixed orientation of the Earth's axis. Indeed, he pre-empted Ridley by explaining magnetically the fixed orientation of all the planets. Resisting Gilbert's magnetic soul, he located the ultimate source of all magnetism in the sphere of the fixed stars.

Johann Kepler's Magnetic Ellipses

No one could have accused Kepler of conservatism as he worked simultaneously on his magnetic cosmology. By Kepler's later account, Gilbert's philosophy was one of the three foundations of his astronomical revolution (the other two were Copernicus' astronomy and Tycho's observations). Kepler's genius moved Gilbert to the centre of the cosmological battleground.

Kepler had already declared his bold form of Copernicanism in his *Cosmographical Mystery* of 1597. The cosmos he described was a real Neoplatonic mystery. He gave the Sun and planets intelligent souls. But he was soon to 'deny that the celestial movements are the work of Mind' and to explore a more mechanical explanation. Undeterred by Gilbert's theory of animate forms, he wrote around 1601 that 'Gilbert the Englishman appears to have made good what was lacking in my arguments'. By January 1603, he thought he could 'demonstrate all the motions of the planets with these same principles'.

By 1605, he had produced his revolutionary physical astronomy. It replaced circular with elliptical orbits and it substituted magnetic forces for solid spheres. Inevitably, when he published in 1609, he was ignored, even by Galileo. He tried again in 1618 with a textbook. Again, there was little response. What finally brought Kepler and magnetic astronomy attention were his monumental *Tabulae Rudolphinae* [*Rudolphine Tables*] of 1627. The tables computed actual planetary positions according to his theory, and they were of unprecedented accuracy. Users of the tables turned to Kepler's theory to find out why they were so good. In Kepler's opinion, it was because he had accomplished 'the transfer of the whole of astronomy from fictitious circles to natural causes'. Modern astronomy was built on Gilbert's magnetic philosophy.

Unlike Gilbert, Kepler combined his new astronomy's interest in physics with the old astronomy's obsession with accuracy. This was a man for whom an error of a mere 8 minutes of arc in the predicted position of Mars had driven him to months of calculations with oddly shaped orbits before he settled on an ellipse. Not for him, then, Stevin's or Ridley's untried assertions that all planets were like Gilbert's Earth. If magnetism was the celestial force, then it had to generate the exact ellipses, not to mention such subtleties as their angle of inclination with the ecliptic.

Kepler seized on the major claim of Book VI of *De Magnete* that magnetism maintained the absolute direction of the Earth's axis. He applied it to all the planets. Like Gilbert, Kepler also treated the Sun as the unique 'prime mover in the cosmos'. Accordingly, it uniquely retained its soul in his new cosmology. Kepler also made the Sun unique in having just one kind of polarity. The Sun's whole surface formed a strange kind of south 'pole', and magnetic virtue emanated from the Sun to form a set of immaterial spokes or radii.

From here on, his system worked mechanically. Some five years before he learned about Galileo's observations of moving sunspots, he had hypothesised that the Sun rotated, sweeping its radii round with it. Since Kepler did not make the planets inertial bodies, he needed these moving radii to push the planets in their almost circular orbits. The power of the push decreased in an inverse proportion (not Newton's inverse square) with distance from the Sun, and this near enough explained his distance law.

Now Kepler could generate his ellipses. Consider the Earth, with its axis magnetically fixed at an angle of 66.5° to its orbital plane. When it is summer in the Northern Hemisphere, the Earth is at 'perihelion' – its furthest distance from the Sun. At the same time, the Earth's North Pole tilts towards the Sun. Kepler

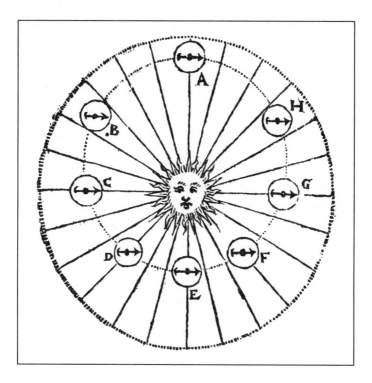

Illustration 17: Kepler's attempt to explain elliptical orbits in terms of planetary magnetism.

The central Sun was a monopole. The planets had inclined but stable magnetic axes like the Earth's. The point of the needle indicates the planetary pole that was attracted by the Sun, and the crotch the repelled pole. At A and E the poles are equidistant from the Sun, and there is no net attraction or repulsion. From A through B, C and D, the attracted pole is closer to the Sun. The planet therefore experienced a net attractive force. It moved closer to the Sun (and also speeded up), reaching its closest approach (aphelion) at E. From E through F, G and H there was a net force of repulsion. The planet receded (and slowed down), reaching perihelion at A. The diagram also shows how Kepler regarded his ellipses as magnetic deviations from a traditional circular orbit.

concluded that the Sun's south monopole attracted the Earth's North Pole just a little more than it repelled the South Pole. Gilbert's principle of the magnetically stabilised axis ensured that the North Pole did not point directly at the Sun.

The Earth therefore experienced a very small net attractive force, which had two tiny effects. The Earth moved closer to the Sun, and began to accelerate. When the Earth progresses in its annual orbit to the autumn equinox, North and South Poles are equidistant from the Sun. Kepler's net attractive force declined to zero. The Earth reached its closest approach to the Sun (aphelion) and obtained maximum speed. As it moves on to the winter solstice, the Earth's South Pole becomes closer to the Sun than its North Pole. For Kepler, that produced a slight net repulsion from the Sun. This acts throughout the second half of the orbit, pushing the Earth away from the Sun and slowing it down.

Kepler's ingenious but incorrect magnetic dynamics gave him what he needed. The Earth's orbit is distorted from a true circle, according to a cosine formula not unlike our (post-Keplerian) equation for an ellipse! The Earth accelerated and decelerated in a satisfactory approximation to his area law. For the other planets, yet to be observed through telescopes that revealed any poles, Kepler could set his own values for the axes' angles of inclination and the strength of their polar attractions and repulsions until the orbits fitted his data.

To account magnetically for other astronomical phenomena, Kepler invented other pairs of poles. One pair explained the different angle of inclination with the ecliptic of each planet's orbital plane. They resulted in a net magnetic push above the ecliptic for half of the orbit and a push below for the other half. After all this, Kepler's magnetic planets had a much more complicated structure than Gilbert's Earth. Opponents accused him of futile speculation. Kepler, a true physical astronomer, thought that the exacting demands of mathematical astronomy 'clears our way to the inward substance of the globes'. If Gilbert's magnetic Copernicanism was not strictly supported by the evidence, then Kepler's was real speculation. But the success of the tables he produced for Emperor Rudolph forced scientists, including pious Catholics, to take it seriously.

Kepler could not publish his works in hard-line Catholic countries bent on enforcing the 1616 decree that the Earth stood still. He had had the misfortune to publish his textbook just as Catholic–Protestant tensions escalated into the Thirty Years War. Several times he had to relocate within the patchwork of central European states to avoid harassment. But only his death in 1630 could save him from a mounting campaign against himself, Gilbert and Stevin.

Magnetism and the Trial of Galileo

Kepler and Gilbert were attacked in 1630 by Libert Froidmont, a conservative theology professor in Louvain, who had decided that the condemnation of 1616 needed defending. He argued that the lodestone's virtue was neither long-range nor universal enough to be a proper celestial force. The year after was the turn of Jean-Baptiste Morin, a Parisian physician, theologian, opponent of Descartes – and developer of a longitude scheme.

He tried to improve on Froidmont's antiquated defence. If Gilbert and Kepler were right, he added, then terrestrial magnetism must be so stupendously strong that no one would be able to lift up a lodestone. Arguments like these were not going to halt the advance of Gilbert's magnetic philosophy.

Nor did they lessen Galileo's opinion that his anti-Copernican opponents were 'mental pygmies'. A combination of Galileo's fame, powerful protectors and his protestations of sincerity kept those opponents at bay until 1632. In that year, he got his *Dialogo sopra i due massimi sistemi del mondo* [*Dialogue Concerning the Two Chief World Systems*] past the Roman censors. The book broke the spirit of his promise to Pope Urban VIII that it would treat geostatic and heliocentric cosmologies even-handedly. It gave his enemies, especially the Jesuits, the chance to bring him to trial for 'vehement suspicion of heresy'.

The Inquisitors presented four proofs of guilt. One was that he 'cites approvingly the opinion of William Gilbert, a perverse and quibbling heretic'. He had, and had probably approved of *De Magnete* since his Venice days in 1602. That was when Sagredo, a friend of Galileo's whom he immortalised as a character in the *Dialogue*, wrote to Gilbert of the 'wonderfull liking' of his book among 'divers learned men of Venice'.

Galileo used two of Gilbert's Copernican arguments in the *Dialogue* – and none of Kepler's. He reproduced Gilbert's demonstration that a magnet performs circular motions and used it, as Gilbert intended, to convince the Aristotelian fall guy in the dialogues that true elemental Earth was magnetic. He also used Gilbert's magnetic explanation of the Earth's axial stability. It is ironic, perhaps, that Galileo's concept of inertia, a concept of circular (not rectilinear) inertia specifically designed to permit Copernican motions, rendered Gilbert's reasons logically redundant. Such was the attraction of Gilbert's philosophy. Not surprisingly, many copies of *De Magnete* have been found with the Copernican sections cut out or mutilated in accordance with Roman censorship laws. Galileo's trial had another consequence. It provoked the Jesuits into some truly brilliant magnetic philosophy.

· CHAPTER 18 ·

ENTER THE JESUITS: THE END OF MAGNETIC COSMOLOGY

Jesuits were charged with taking on natural philosophy that threatened Aquinas' synthesis of Aristotelianism and Christianity. Gilbert and the Jesuits were therefore natural enemies. Their most effective strategy was not blind opposition, but clever incorporation. So they took on the apparently impossible task of bringing Gilbert's magnetic philosophy into line with Aristotelianism.

Niccolo Cabeo's Aristotelian Magnets

The breakthrough was made by the Italian Niccolo Cabeo, a philosopher whose career and cleverness earned him the sobriquet 'the Jesuit Galileo'. He published his *Philosophia Magnetica* [*Magnetic Philosophy*] in 1629. Cabeo was convinced by Gilbert's proofs of the magnetic Earth. Of course, he could not accept that the Earth moved. With Galileo's trial still four years ahead, he seemed to regard Copernicanism as a minor irritant. Proponents like Gilbert were 'so clearly in error that I need not dwell on them'. But he was provoked by Gilbert's gleeful conclusion that Aristotelian matter theory was fundamentally wrong. On the contrary, he proclaimed; what Gilbert had really discovered was a new Aristotelian prime quality that philosophers had overlooked for 2,000 years.

Rather like Mendeleev's Periodic Table led chemists to

discover new elements to fill the holes, Cabeo argued that magnetism fitted perfectly into holes in the Aristotelian classification of qualities. This is where our earlier discussion of Aristotelian distinctions in chapter 3 becomes important. You remember (of course you do!) that the six prime Aristotelian qualities could be divided into alterative ones (heat and moistness), locomotive ones (gravity and levity) and passive ones (cold and dryness).

Cabeo pointed out that logic required a fourth category for qualities that both alter and move substances. No examples had been identified, but was this not what magnetism did? When unmagnetised iron was brought near a lodestone, it was altered (from being non-magnetic to magnetic) and then moved by attraction. St Thomas had, in fact, said something very like this.

Cabeo then concentrated upon the locomotions that magnetism produced by its nature. Aristotelians had classified motions according to privileged points in the universe. There were motions towards, away from and around the centre. Cabeo pointed out there are two other privileged places in the universe. These are the heavenly poles that are joined together by the giant celestial axis that cut through the Earth. Aristotelians had not previously thought of motion to the poles as a fundamental natural kind, but, said Cabeo, they should have done. Gilbert's analysis of magnetic motion proved just how fundamental it was.

Cabeo reflected again on the Aristotelian kinds of locomotion. There were circular and rectilinear motions. Rectilinear came in two kinds. The motions of earth and fire were complete, water and air incomplete. You've probably anticipated Cabeo by now. Logically, there was a fourth fundamental kind – incomplete circular motions. Gilbert had discovered their existence too.

Cabeo had demonstrated that magnetism was a new, prime Aristotelian quality. His reworking of Gilbert's magnetic philosophy for his own purposes was as compelling in its own way as Kepler's. Within thirty years, two things that Gilbert thought impossible had happened. Mathematics and natural philosophy had combined to make magnetic virtue the foundation of a new physical astronomy. Aristotelians had defended their theory of matter and captured magnetism for the forces of reaction.

The history of science offers few better examples of the fact that experiments alone prove nothing, despite our celebration of Gilbert as the first scientific author to make experiments his best kind of proof. In *Magnetic Philosophy*, Cabeo accepted and reproduced almost all of Gilbert's experiments, and included his own 'improved' version of the deformed terrella. But he denied that Gilbert's experiments actually demonstrated anything about the cause of magnetism. Gilbert had realised that magnetism was cosmically crucial, not a typical occult quality. He had just been too prejudiced to realise that it was actually a new Aristotelian prime quality, present in the form of elemental earth alongside cold, dryness and gravity.

Cabeo's causal 'demonstration' turned magnetic philosophy into an Aristotelian science. And if magnetism was an Aristotelian quality, then it certainly didn't move the Earth. Quite the opposite. Aristotelian qualities only moved things in order to bring them to rest in their natural place. Once the magnetic quality had turned an earthy body into alignment with the poles its work was done. The piece of earth was at rest.

Athanasius Kircher's Zealous Magnets

In the aftermath of Galileo's trial, Cabeo's Jesuit colleagues pressed home the anti-Copernican consequences of a magnetic Aristotelian Earth. The most zealous was Athanasius Kircher, who published more pages of magnetic philosophy than anyone in history. Driven like Kepler out of Germany, in his case by Protestant forces, the extraordinary polymath was promoted to the Jesuits' Roman College in 1634, and he immediately took a hard line against Copernicans.

Magnetism was an ideal subject for Kircher. Like his English equivalent, Robert Fludd, his world view mixed ancient and modern natural philosophy. He believed that the universe was full of occult virtues and magical sympathies, of which magnetism was a prime example. Gilbert thought that he had rescued magnetism from occultism. *De Magnete* simply allowed Kircher and Fludd to write at greater length on the magnetic foundation of magic, to much less public acclaim than Cabeo or Kepler, it must be said.

Kircher's huge book *Magnes* [*The Lodestone*] was an encyclopaedic mixture of contemporary magnetic experiments and ancient wisdom. He devoted a whole book to magnetic cosmology. In it he attacked the 'heretics' Gilbert, Stevin and Kepler and posed the question 'Does a magnetic force really exist in the Earth, the Sun and

Illustration 18: Title page of Athanasius Kircher,
Magnes sive de Arte Magnetica (Rome, 1641).
Kircher's voluminous and lavish work *The Lodestone, or the Magnetic Art* was one of several on magnetic philosophy funded by his order, the Society of Jesus. The Catholic Kircher wrote more on magnetism than any other author, and tried to refute magnetic Copernicanism. The images exemplify Kircher's encyclopaedic, eclectic and occultist treatment of magnetism and many other subjects.

other wandering and fixed stars, and do they really and properly attract each other magnetically?' The right answer was urgent.

And, as it was my duty, I wanted to investigate this trifling opinion because of my zeal for the honour of God, the Holy Mother and the Church (especially since I know that no-one has publicly disproved these magnetic motions of the heavenly bodies). This [opinion] *is not only so destructive to the Christian Republic, but it is also dangerous to faith.*

Gilbert was wrong to conclude that the Earth moved magnetically. 'He should rather have defended the opposite, since there is nothing more certain in all Magnetic Philosophy, which more certainly and truly confirms the Earth's stability.' Kircher did no more than draw out the implications of Cabeo's work. The Earth's magnetic quality brings it to rest. As such, it actively resists continual rotation.

As for Kepler's speculations, Kircher tested them experimentally. He had a model solar system built for his famous museum. The planets were glass balls containing magnets, which ran in circular grooves. 'In the centre is placed a lodestone of the most exquisite power, which denotes Kepler's solar magnet.' Kircher demonstrated his 'ocular confutation' to his many visitors, cheerfully pointing out that although the glass planets orbited his 'sun', they neither rotated on their axes nor obeyed Kepler's distance law. It would have been unhelpful to point out that Kepler's celestial magnets were not normal lodestones.

Illustration 19: Title page of Jacques Grandami, *Nova Demonstratio Immobilitatis Terrae* (La Flèche, 1645). In the centre a cherub performs Grandami's crucial experiment: when floated, a vertically mounted terrella always comes to rest in the same east–west alignment, suggesting that magnetism actually prevents the Earth rotating. Contemporaries took the experiment very seriously. The Biblical verse at the top, Ecclesiastes I:4,

Jacques Grandami's Amazing Magnets

Jacques Grandami, S.J., followed in the same tradition. He was the epitome of the progressive, liberal Jesuit. Away from Rome, in the French Collège de la Flèche that Descartes attended, Grandami's own Jesuit professor had taught the inertial mechanics of Galileo. The debate had moved on. Jesuits needed what Grandami advertised in his title, *A New Demonstration of the Earth's immobility*.

The problem with the old demonstration was that it still relied upon gravity. The Earth could not move from its location because it would require an unnaturally powerful agent to heave the Earth's entire bulk from its resting place at the centre of the universe. That ruled out full Copernicanism, but what about the diurnal rotation, upon which *De Magnete* focused? If the Earth merely spun on its axis, then none of its parts were moving any further from, or closer to, the centre, and gravity would have no stabilising effect. The Jesuit philosopher Biancarni had produced a counter-argument. But if Galileo was right about inertia, then Biancarni's defence did not work.

In the *New Demonstration*, Grandami happily conceded that Galileo was indeed right.

Gravity only impedes upward, downward or lateral rectilinear motion. In no way does it impede circular

reads 'Terra in aeturnum stat' [the Earth endures (stands) for ever]. Grandami, a Jesuit, believed his magnetic experiment proved that recent Catholic theologians had been right to interpret the verse literally. Notice also the references to a universal true magnetic meridian and finding longitude. Another cherub holds Grandami's supposed longitude-finding compass.

motion about a centre in the same place. This is clearly understood from a consideration of all heavy bodies, which move with great ease about their centre or poles when they are suspended in air, floated on water or even when stood on a flat polished surface. . . . So if the earth had no other quality besides gravity, then, although it would always be in the centre of the elemental world, it could easily move around that centre.

Grandami could make his nonchalant concession because, thanks to Gilbert, Cabeo and his own twenty years of work, he could imbue the Earth with gravity *and* magnetism.

The 'new philosophers' of the mid-seventeenth century had tended to ignore the theories of Cabeo and Kircher. They were what one expected from Jesuits. (One might add that crazy ideas about screw-threaded magnetic particles were what one expected from mechanistic philosophers.) But Grandami's work could not be ignored. In the *New Demonstration* he produced a stunning experimental ace straight out of Gilbert's pack.

Take a terrella and float it on a wooden boat, as Gilbert described. But take pains to make sure that the terrella's axis is absolutely vertical. Mark the terrella's equator with the 180° of longitude east and west. Now spin the boat around. The terrella is spinning on its axis just like Gilbert's Earth. Of course, it doesn't spin forever but comes to rest. Note carefully the orientation of the equator when it stops. Spin it again, note the orientation, and repeat.

Grandami had found that his terrella, when arranged this way, always came to rest in the same East–West orientation! He had 'discovered' that magnetism controlled not only North–South alignments, but East–West too! His conclusion? Although modern theories of gravity showed that it could not prevent diurnal rotation, 'there is no doubt that the magnetic virtue that God gave to [the Earth] not only keeps its poles still and stable but also its other parts and points'.

This was his new demonstration. Armed with it, people would realise that magnetic Copernicanism ...

> ... [was] *not only false, but clearly contrary to magnetic laws. They will have our demonstration, stick to their opinions more firmly, and refute opposing opinions more solidly. Then they can praise the divine wisdom in the earth's magnetic quality, which causes its stability, and demolish the other useless and ridiculous effects of the Sun and other planets.*

Grandami circulated his extraordinary experimental claim in a manuscript in 1641. As usual, Mersenne sent copies to his correspondents. Amazingly, it was accepted as a new magnetic fact, reproduced in numerous authoritative treatises. Descartes included it in *Principles of Philosophy* before Grandami published it himself. Later Jesuit magneticians such as Zucchi, Schott and Léotaud obviously gave it prominence. Those who tried to replicate it reported that, broadly speaking, Grandami was right. In the 1670s, Leibniz pestered the Royal Society for its results, but there were none. Modern magnetohydrodynamicists of my acquaintance have not replicated it, but they think the experiment would work.

Had Grandami trumped Gilbert's Copernican magnetic Earth with a geostatic one? The logic that had turned magnetism into a cosmic principle seemed to demand it. His demonstration conformed with Gilbert's central principle of analogy: the terrella is a model Earth that reproduces the phenomena of the giant lodestone.

Unfortunately for Grandami, no one outside La Flèche College wanted to accept the logic of his demonstration. Jesuit colleagues judged him too radical. He had conceded too much to the emerging inertial theories of Galileo and Descartes. Moreover, he had transgressed the new Aristotelian principles set out by Cabeo – all magnetic motion was directed towards the poles, and Grandami's wasn't.

Copernicans did not even begin to wonder whether Grandami had routed Gilbert and Kepler. The most interesting and significant response came from the neutral corner of Marin Mersenne. The Minim friar's philosophical scepticism had for some time led him into relativism. He could see that Gilbert, Galileo, Kepler and Descartes had developed a physics of a moving Earth that plausibly countered Aristotelian arguments. He thought that the two sides had reached stalemate; the Copernican question was physically undecidable. The liberal friar also thought it was theologically undecidable, and he especially disliked the Jesuits' dogmatism. He was, he said, tired of work like Grandami's. It may have been tiresome, but was it wrong? Mersenne was sure that it was.

The End of the Little Earth

As Mersenne tried to pick a hole in Grandami's reasoning, he rediscovered a devastating flaw that would collapse the whole edifice of magnetic cosmology. His objection was virtually the same as that made by William Barlow against Ridley's Copernicanism. Faced with the Gilbertmania of the early seventeenth century, Barlow's bomb had failed to go off. When the Jesuits stepped into the Gilbertians' territory, it blew both camps to bits.

Mersenne reasoned that there was a fundamental disanalogy between terrella and terra. The terrella lay inside the Earth's magnetic sphere of influence. Like all small magnets, its motions were controlled, over-powered even, by a much larger one – in this case, the Earth. The Earth was a free magnet, or (if one wanted to accommodate Kepler and his odd solar magnet) con-trolled in a different way. In his own notes to Grandami's manuscript, Mersenne wrote: 'The reason he produces for the orientation of this lodestone with the Earth, namely that its parts are analogues with the parts of the Earth from which they draw their power, is not adequate.'

Mersenne committed his view to print in 1644 in a pre-emptive strike against Grandami and the Jesuits' magnetic Earth. It was his final statement on the Coper-nican question.

I do not even wish to pursue those ways by which others believe that the Earth's stability is proved from magnetic directions, since exactly the same thing happens to the magnet whether the Earth stands still or moves. So far, nothing from magnetic laws, any more than from projectiles and the fall of heavy bodies can or ought to prove the double or triple motion of the Earth, the double rest of the Earth or its immobility.

As other new philosophers came to grapple with Grandami's experiment, they adopted similar positions. In 1663, the Halifax doctor, Fellow of the Royal Society and mechanical philosopher Henry Power published some *Experiments Magneticall: with a Confutation of Grandamicus*. He had composed an entire manuscript, 'Magneticall Philosophy', the first by an Englishman since Ridley and Barlow, but the Royal Society was most interested in its novel experiments and well-aimed boot at the Jesuit. Like many others, Power accepted Grandami's experiment, but not his explanation.

> *We answer, That the reason why the Terrella does wheel about, and direct certain parts of its Aequator to certain and determinate points of the Horizon, is, that it is overpowered by the Magnetic Effluxions of the Earth; which, as a greater Magnet, does violently reduce it to that Situation* [in which it was formed]: *And therefore this great Argument against the Dineticall Motion of the Earth, is no Argument at all, unless he could prove to us that the Terrella could play this trick;* [if] *it were removed out of the sphaere of the Earth's Magnetism?, which is beyond his Philosophy ever to demonstrate.*

Power thought that he had saved his countryman's magnetic cosmology from Jesuitry. In fact, his solution showed that the glory days of magnetic philosophy were over, even in England. In a way that he could not have expected, Grandami had ended the reign of magnetism in the Copernican universe.

CHAPTER 19

THE LONGITUDE FOUND,
AND LOST AGAIN

Magnetic cosmology was an extraordinary episode in the history of magnetism. It makes Gilbert's decision to leave out from *De Magnete* any motions 'to which we cannot assign with certainty any natural causes' look like a model of cautious empiricism. The collapse did not of itself bring an end to magnetic philosophy. If magnetic planets were speculative, magnetic navigation was not.

By 1650, magnetic navigation was in trouble too. Magnetic latitudes derived from inclination had not proved useful, and 'Dr Gilbert's rule' of the variation did not seem to work reliably. But deep trouble began with the discovery of secular variation in London in 1634.

London's Gilbertian research programme in magnetic navigation survived the death of Prince Henry. The new centre was Gresham College, founded in 1597 to promote useful learning, including geometry and its practical applications. Close links existed between Gresham professors who knew their *De Magnete* and naval personnel and instrument makers. Gilbert's philosophy provided theoretical underpinnings for the English belief that variation was irregular and needed to be observed, not calculated as de Nautonnier had done.

Observations in the naval dockyards of east London were very accurate. Naval employees brought their best needles and lodestones, while Gresham professors brought their astronomical and mathematical precision. William Borough had provided the 'standard' value for London of 11°16'E in 1581, which *De Magnete* circulated to the wider world. When the group got a value near 6°E in 1622 they thought little of it, but results of 4°E in 1633 and 1634 caused a minor panic. Was it an error, the result of the hidden variables that plagued variation data? After several tests, Professor Henry Gellibrand decided it was not. His group had used the same needles, stones, quadrants and methods, in the very same place, as in 1622.

The 'Variation of the Variation'

Gellibrand announced the 'diminution' or 'variation of the variation' in his *Discourse Mathematical on the variation of the Needle* of 1635. We now know that he was the first person in over 100 years of careful measurements by sailors and scholars to have consciously observed secular variation. Today, we think of it as the global change over time caused by cyclical changes in the geodynamo. The Earth's tilted dipole wobbles, and the magnetic poles wander with it. It is almost the magnetic equivalent of the precession of the Earth's axis of rotation.

Gellibrand seriously wondered whether there *had* been a shift of the Earth's axis. The explanation appealed to him. His *Discourse* shows that he was still a Gilbertian. He did not believe that the magnetic poles were separate from the geographical poles, and he didn't conclude that they had motion of their own.

Predictably, it was Mersenne who circulated Gellibrand's claim to the disbelieving continent of Europe. Until confirmations accumulated in the early 1640s, many people thought that the unknown Englishman had been deceived.

Once it was confirmed, Gellibrand's discovery inaugurated the final wave of magnetic longitude and latitude schemes. If variation was not time-invariant then Gilbert's theory was wrong. There was now another explanation for the apparent irregularities. All previous variation measurements, few of which could be dated, were unreliable. Maybe compass needles had, after all, always pointed directly to the magnetic poles. De Nautonnier's tilted dipoles were back, only this time they were moving. Gilbert's magnetic philosophy was in retreat, even in his homeland.

Henry Bond and his Moving Poles

The first magnetic longitude and latitude scheme based on a tilted, moving dipole model was devised in 1637 by the London navigation lecturer, Henry Bond. The model was 'mecometrically' useful only if needles pointed directly to the poles, only if variation was as utterly regular as de Nautonnier had thought in 1602. Bond's scheme was basically de Nautonnier's with a temporal component. He ignored evidence of irregularity, just as he ignored Gilbert's still sound reasoning that the magnetic Earth was not perfect.

Continental hopefuls came and went with similar schemes. Bond refined and promoted the scheme throughout his long life. He surprised fellows of London's Royal Society with his successful prediction of the variation which, after centuries of 'northeasting', now northwested. With powerful backers, he presented his solution of the longitude problem to a special royal committee in 1674. The panel prevaricated, but it didn't buy it. Bond resorted to publishing. *The Longitude Found* appeared in 1676, cruelly answered by *The Longitude Not Found*. Bond died a poor, bitter man.

The new schemes soon foundered on the same rock that had sunk the Portuguese prime meridian theorists, de Nautonnier and their numerous imitators. Even allowing for its secular component, variation just wasn't regular enough. In a sense, Gilbert was partially vindicated. Divine wisdom may have implanted magnetic virtue in the Earth, but when it came to magnetic navigation, God was not a regular guy.

Grandami's Last Stand

This returns us to Grandami. The Jesuit's God was regular, a providential God who had given mariners the magnetic tools they needed. Besides the non-rotating terrella, the *New Demonstration* had another ace up its sleeve – the Renaissance mariner's dream of a non-varying compass needle that pointed true north and identified the true magnetic meridians needed for the longitude.

He was led to it by his cosmological convictions. Jesuit magnetic philosophy depended just as much as Gilbert's upon the identity and stability of the Earth's magnetic and geographical poles. Not for him, then, the renewed fashion for tilted dipoles, nor the likelihood of secular variation. He declared that Gellibrand must have been incompetent. His judgement was plausible in 1641 when he circulated his manuscript. It was not when he published it unaltered in 1645. And his needle was a joke. Even Jesuits were appalled by his intervention in the practical art of navigation.

With Bond, Grandami and the other schemers, a chapter closes in the history of the magnetic Earth. Gilbert had cemented nearly a century of confidence in the potential of magnetism to revolutionise navigation, a confidence that would never again be widely shared. Initially, it seemed that *De Magnete* had eliminated, or at least explained, the disorderly nature of geomagnetic data. Gilbert inspired a series of empirical and theoretical developments that led, by turns, to understanding, hopes for the longitude, fatal anomalies, new understandings and hopes. In the end, confronted by the complexity of geomagnetism, the investigators despaired. It is an emotion equally familiar to modern geomagneticians. After Bond, the focus of the story moves from *Latitude & the Magnetic Earth* and Gilbert's magnetism to Dava Sobel's *Longitude* and Harrison's clocks.

CONCLUSION: IT'S THE END OF THE WORLD AS WE KNOW IT

Henry Power's work of 'Magneticall Philosophy' was the last English treatise on magnetism for several decades, and the last to ally itself closely with the tradition established by Gilbert's *De Magnete*. Despite its title, Power's book illustrated many of the causes of the end of magnetic philosophy. First, like Barlow and Mersenne, he had abandoned Gilbert's central principle of analogy. Without its function as an exact model of the Earth, the terrella lost its significance. After Grandami, no one could argue that laboratory results with a terrella and versorium illuminated the true nature of geomagnetism.

Just as destructive were Power's references to an effluvial or mechanical theory of magnetism. Power's theory came from Descartes, and by 1663 most new philosophers had adopted some version of it. Few people had ever been at ease with Gilbert's immaterialist account of magnetic virtue. Not all mechanical philosophers were convinced by Descartes' ingenious theory, but they were sure that magnetism was mediated by some emission or 'effluvium' of sub-microscopic particles. This was how mechanical philosophers of all stripes, followers of Descartes, Gassendi, Hobbes or the cautious Robert Boyle, explained a range of phenomena. Sound was transmitted by air particles; climate depended on watery vapours; odours were atomic evaporations from smelly substances; light was most likely caused by a stream of particles; heat was the rapid motion of any particles.

And so too was magnetism. To say otherwise was to re-admit occult qualities and sympathies to the realm of science. In the process of mechanising magnetism, philosophers like Descartes, Boyle and Henry Power effectively denied it a special status, denied that it was a fundamental force in the universe. The science of magnetism persisted, Gilbert-style experiments contin-ued, the Earth remained magnetic, but Gilbert's mag-netic philosophy of the Earth was dead. Magnetism was weird and important, but it was not a fundamental property of nature. Ultimately, it was just another stream of colliding particles, and it was not the motor of terrestrial or other heavenly motions.

The rise of mechanical philosophy explains why Henry Power and his numerous fellow Copernicans could treat Gilbert's central principle literally with abandon. For the same reason as they preferred materialist expla-nations of the lodestone, most philosophers preferred Descartes' or similarly mechanical accounts of the solar system. The Earth spun inertially and was pushed around the Sun. Copernican philosophers didn't need to rely on Gilbert's magnetic philosophy any more. Grandami had unwittingly helped them to realise it.

Grandami's illusory needle played its part in the fourth reason for the decline. The erosion of confidence in magnetic longitude and latitude schemes diminished the practical value of magnetic philosophy. And, even as they faded, they encouraged natural philosophers to accept the reality of tilted dipoles. Gilbert's *De Magnete* held nothing like the interest in 1670 that it had commanded in the decades after 1600. Magnetic philosophy retreated from the centre of the scientific world. Shorn of its Earth-shattering potential, it was transformed from 'a new philosophy' into the dull but worthy book 'on the magnet' that has been celebrated by scientific empiricists since Victorian times.

The end of magnetic philosophy was not, of course, the end of the magnetic Earth or magnetic science. We return to the modern world by retracing the steps of the introduction. Newton turned magnetism back into a non-mechanical principle of matter, albeit not as important as gravity. The early Newtonians restored a little fashionableness to magnetism, and one even toyed with a magnetic longitude scheme. Compasses were continually improved. By the late eighteenth century, fluid theories were so in vogue that Mesmer's animal magnetic fluid seemed plausible. And then, in the nineteenth century, magnetism was rejoined to Gilbert's neglected child of electricity. The rest, too, is history.

The irony is that none of the factors that stimulated Gilbert to create the magnetic Earth and sustained his 'magnetic philosophy' have any relevance today. References to magnetic souls are confined to the wackiest fringes of the World Wide Web. The Earth has long since moved by other means. Modern navigators have global positioning systems that point to satellites, not to the magnetic poles. The cliché about the past being another country is as true of the history of magnetism as of anything else. On their way to the New World, William Gilbert and magnetic navigators did things differently there.

FURTHER READING

Gilbert Himself

I hope you'll agree that *De Magnete* is a good read. It has been translated into English twice. The better edition is *On the Magnet, Magnetick Bodies also, and on the great magnet, the earth*, translated by Sylvanus P. Thompson (London, 1900; reprint edn, New York, 1958). Originally an edition to mark the tercentenary of *De Magnete*, it preserves the layout of the Latin original and has useful notes, but it is expensive and rare. I have therefore quoted from the common edition, *De Magnete*, translated by P. Fleury Mottelay (New York, 1893; reprint edn, New York, 1958). It is published in paperback by Dover Publications, and is widely available at about US$14.00. In both translations, readers should watch out for modern-looking terms such as 'force'. Gilbert's Latin terms (such as *vis*, which is better rendered as 'power') do not translate easily on to those of modern science.

Gilbert's 'other book' is *De Mundo nostro Sublunari Philosophia Nova* (Amsterdam, 1651; reprint edn, Amsterdam, 1965). You need to be able to read Latin and have access to a very good library. My colleague Ian Stewart and I plan to publish a translation in 2004.

Gilbert's sources are not widely available. There are several editions of Nicolas Copernicus, *On the Revolutions of the Heavenly Spheres*, of which Book I can be enjoyed by the non-specialist. Edward Rosen's excellent translation is available on the Internet at www.dartmouth.edu/~matc/readers/renaissance.astro/1.1.Revol.html

The book on the lodestone in Giambattista della Porta's *Natural Magick* is also on the net at http://members.tscnet.com/pages/omard1/jportac7.html (be warned that it is an unreliable English translation of 1658).

General Reading

The scientific background to Gilbert's revolution is best covered in David C. Lindberg, *The Beginnings of Western Science: The European Scientific Tradition in Philosophical, Religious, and Institutional Context, 600 B.C. to A.D. 1450* (Chicago: University of Chicago Press, 1992). The sixteenth-century intellectual context to Gilbert's thought is nicely treated by Brian Copenhaver and Charles Schmitt, *Renaissance Philosophy* (Oxford: Oxford University Press, 1991), volume III of the series *A History of Western Philosophy*. The changes in natural philosophy to which Gilbert contributed have been given a very up-to-date treatment in Peter Dear, *Revolutionising the Sciences: European Knowledge and its Ambitions 1500–1700* (Basingstoke: Palgrave, 2001).

The seventeenth century gets lively coverage in Steven Shapin, *The Scientific Revolution* (Chicago: Chicago University Press, 1996). An excellent short survey is John Henry, *The Scientific Revolution and the Origins of Modern Science* (Basingstoke: Macmillan, 1997). John Henry has also published, in this series, *Moving Heaven and Earth: Copernicus and the Solar System* (Cambridge: Icon Books,

2001). The general history of navigation has not received so much recent attention. E. G. R. Taylor, *The Haven-finding Art. A History of Navigation from Odysseus to Captain Cook* (London: Hollis and Carter, 1956) remains a good introduction.

The History of Mathematics and Navigation

Enthusiasts of navigation will prefer to Taylor *The Art of Navigation in England in Elizabethan and Early Stuart Times* by David W. Waters (London: Hollis and Carter, 1958). The definitive survey of the theory and practice of specifically *magnetic* navigation has just appeared. This is the two-volume work by A. R. T. Jonkers, *North by Northwest. Seafaring, Science and the Earth's Magnetic Field (1600–1800)* (Gottingen: Cuvillier Verlag, no date given). It also contains almost all one needs to know about current geomagnetic theory and models. A version should be published by Johns Hopkins University Press in 2003. A valuable history of English mathematics in Gilbert's time, which takes a more positive view than Gilbert (and me) about the universities, is Mordechai Feingold, *The Mathematicians' Apprenticeship: Science, Universities and Society in England, 1560–1640* (Cambridge: Cambridge University Press, 1984).

Books on Gilbert

The first work of modern scholarship on Gilbert was Duane H. D. Roller, *The De Magnete of William Gilbert* (Amsterdam: Menno Hertzberger, 1959). A very good English account of *De Mundo* is Sister Suzanne Kelly, *The*

De Mundo of William Gilbert (Amsterdam: Menno Hertzberger, 1965). *Latitude & the Magnetic Earth* is the first subsequent book to focus on Gilbert.

Selected Scholarly Articles on Gilbert

Gilbert's relation to medieval scholasticism was discussed by W. James King, 'The natural philosophy of William Gilbert and his predecessors', *United States National Museum Bulletin* vol. 218 (1959), pp. 121–39. His debt to craftsmen was explored by the Marxist Edgar Zilsel in 'The origins of William Gilbert's scientific method', *Journal of the History of Ideas*, vol. 2 (1941), pp. 1–32. The importance of mathematical practitioners is the subject of J. A. Bennett, 'The mechanics' philosophy and the mechanical philosophy', *History of Science*, vol. 24 (1986), pp. 1–28. Bennett has also written on 'Cosmology and the magnetic philosophy, 1640–1680', *Journal of the History of Astronomy*, vol. 12 (1981), pp. 165–77. On Jesuit cosmology, see Martha Baldwin, 'Magnetism and the anti-Copernican polemic', *Journal of the History of Astronomy*, vol. 16 (1985), pp. 155–74. Another important contribution was Gad Freudenthal, 'Theory of matter and cosmology in William Gilbert's *De Magnete*', *Isis*, vol. 74 (1983), pp. 22–37. An important article is John Henry, 'Animism and empiricism: Copernican physics and the origins of William Gilbert's experimental method', *Journal of the History of Ideas*, vol. 62 (2001), pp. 99–119. Another recent work of interest is Eileen Reeves, 'Old wives' tales and the new world system: Gilbert, Galileo, and Kepler', *Configurations*, vol. 7 (1999), pp. 301–35.

Immodest Suggestions

A book of this kind generally does not include footnotes or an extensive bibliography. Readers frustrated by my lack of citations will find some relief in my more scholarly publications. I discussed Gilbert's history of science in 'The history of science and the Renaissance science of history', in Stephen Pumfrey, Paolo L. Rossi and Maurice Slawinski (eds), *Science, Culture and Popular Belief in Renaissance Europe* (Manchester: Manchester University Press, 1991), pp. 48–70. For the Aristotelian response to Gilbert, see 'Neo-Aristotelianism and the magnetic philosophy', in John Henry and Sarah Hutton (eds), *New Perspectives on Renaissance Thought: Essays in the History of Science, Education and Philosophy* (Duckworth: London, 1990), pp. 177–89. For Gilbert and theories of variation see '"O tempora, O magnes!" A sociological analysis of the discovery of secular magnetic variation in 1634', *British Journal for the History of Science*, vol. 22 (1989), pp. 181–214. For Kepler and other magnetic astronomers, see 'Magnetical philosophy and astronomy, 1600–1650', in R. Taton and C. Wilson (eds), *The General History of Astronomy, vol. 2, part A* (Cambridge: Cambridge University Press, 1989), pp. 45–53. For Gilbert's life, see the extensive article 'Gilbert, William', in *New Dictionary of National Biography* (Oxford: Oxford University Press, forthcoming). I am happy to supply further evidence upon reasonable request via the William Gilbert website at www.lancs.ac.uk/depts/history/histwebsite/whatson/research/gilbert.htm

In case of difficulty in obtaining any Icon title through normal channels, books can be purchased through BOOKPOST.

Tel: + 44 1624 836000
Fax: + 44 1624 837033
E-mail: bookshop@enterprise.net
www.bookpost.co.uk

Please quote 'Ref: Faber' when placing your order.

If you require further assistance, please contact:
info@iconbooks.co.uk